Mathematical Foundations of Statistical Mechanics: Algorithms and Computations

About the Book

This book aims to predict the Mathematical Foundations of Statistical Mechanics and explain the measurable properties of macroscopic systems on the basis of the properties and behaviour of the microscopic constituents of those systems.

Entropy of an isolated system may increase spontaneously into the future but not into the past. But the dynamical laws governing the motion of the micro-constituents are, at least on the standard views of those laws as being the usual laws of classical or quantum dynamics, time reversal invariant. Introducing probabilistic elements into the underlying theory still does not by itself explain where time asymmetry gets into the explanatory account.

About the Author

Dr. K.K Tiwari is Principal at Takshshila College, Rachna Nagar, Bhopal. Previously he was Professor & Head, Department of Computer Science at same college. He is also M.P. Governor Nominee Executive Council Member at Devi Ahilya University, Indore (NAAC "A" Grade). His several research papers have been published in reputed computer science journals.

Mathematical Foundations of Statistical Mechanics: Algorithms and Computations

Dr. K.K. Tiwari

AMIGA PRESS INC.

DELHI (INDIA)

AMIGA PRESS INC.
H.O.: 28-A, Indra Park, Chander Nagar
Delhi-110051 (India)
Tel.: 011-22504028, +91-98736-57000
E-mail: amigapress@gmail.com
amigapub@gmail.com
Website : www.amigapress.co.in

This edition is published by Amiga Press Inc.

ISBN: 978-93-87295-35-3

First Published 2019

Printed and bound by Replika Press India Pvt. Ltd.

Preface

Statistical mechanics was the first foundational physical theory in which probabilistic concepts and probabilistic explanation played a fundamental role. Statistical mechanics, branch of physics that combines the principles and procedures of statistics with the laws of both classical and quantum mechanics, particularly with respect to the field of thermodynamics.

It aims to predict and explain the measurable properties of macroscopic systems on the basis of the properties and behaviour of the microscopic constituents of those systems.

The thermodynamic principles demand a world in which physical processes are asymmetric in time. Entropy of an isolated system may increase spontaneously into the future but not into the past. But the dynamical laws governing the motion of the micro-constituents are, at least on the standard views of those laws as being the usual laws of classical or quantum dynamics, time reversal invariant. Introducing probabilistic elements into the underlying theory still does not by itself explain where time asymmetry gets into the explanatory account. Even if, following Maxwell, we take the Second Law of thermodynamics to be merely probabilistic in its assertions, it remains time asymmetric.

Thermodynamics describes a large class of phenomena we observe in macroscopic systems. The aim of statistical mechanics is to account for this behaviour in terms of the dynamical laws governing the microscopic constituents of macroscopic systems and probabilistic assumptions. This article provides a survey of the discussion about the foundation of statistical mechanics by introducing the basic approaches and discussing their merits as well as their problems.

After a brief review of classical mechanics, which provides the background against which statistical mechanics is formulated, we discuss the two main theoretical approaches to statistical mechanics, one of which can be associated with Boltzmann and the other with Gibbs.

The contribution to this subject made over the last decade by researchers applying the techniques of statistical mechanics is the subject of this book.

– Author

Contents

1

Mathemactical Foundations

The kinetic theory of matter attempts to explain the empirical regularities occurring in the macroscopic properties of material objects in terms of the microscopic behaviour of their atomic and molecular constituents as described by Newton's laws of mechanics.

The early studies in the revival of this theory which occurred in the second half of the nineteenth century had only limited explanatory power since they assumed that:

- These constituents all had the same velocity; and
- The influence of intermolecular forces and impacts were negligible. The development of the subject took a major leap forward when Maxwell allowed intermolecular collisions to play a role and demonstrated that their effect was to produce a statistical distribution of molecular velocities in which all velocities would occur with a known probability. This is given by Maxwell's famous distribution law:

$$f(x) = \frac{1}{a\sqrt{\pi}} e^{-x^2/a2}$$

where a is a constant and x is the component of the velocity in the x-direction.

Of course, the statistical element was introduced purely as a matter of convenience in order to overcome the practical problems in handling enormously large numbers of atoms and molecules. The latter themselves were regarded as traversing distinct, continuous space-time trajectories and as obeying Newton's laws.

The original proof of the above law was widely regarded as unsatisfactory and subsequent attempts to derive it more rigorously can be divided into three types:

1. If a Maxwellian distribution is already established then it can be shown that conservation of energy implies that further collisions will leave it unchanged. Hence this distribution is the only one which is stable. This was the approach taken by Maxwell himself.
2. One may define a quantity, *H*, which depends on the velocity distribution and then show that the effect of molecular collisions is always to decrease *H*, unless the distribution is Maxwellian, in which case *H* remains constant. This constitutes the essence of Boltzmann's '*H*-Theorem' approach.
3. One may regard the atomic and molecular velocities as random quantities and obtain the best possible estimate of their distribution, subject to the constraints of fixed energy and number of atoms or molecules, using probability theory. Maxwell's derivation may be improved by calculating all the possible ways of dividing the energy among the atoms or molecules, subject to the above constraints. This was the thinking which subsequently developed into the 'Combinatorial Approach' (and which lies behind the permutation argument above).

Approaches II and III rest on very different foundations: the H-Theorem Approach is explicitly based on a consideration of molecular trajectories, whereas the Combinatorial Approach eschewed such detailed consideration and was simply concerned with the distribution of molecules over energy states. Historically, the Combinatorial Approach played an absolutely fundamental role in the development of quantum theory.

Maxwell's work had an enormous impact upon Boltzmann, who was seeking a mechanical explanation of the apparent irreversibility of natural processes as expressed by the Second Law of Thermodynamics. Put rather crudely, this states that it is impossible for heat to flow from a colder to a hotter body. The law was expressed by Clausius in 1865 in terms of his newly introduced entropy function S, where,

$$dQ < \mathrm{T}.\, dS$$

where dQ is the heat change, T is the temperature change and dS is the change in entropy. The equality holds for reversible processes, of course. The task facing Boltzmann was then two-fold: first, to demonstrate the existence of an entropy function satisfying Clausius's definition and secondly to show that this function could only increase in an irreversible process. After reading Maxwell's 1867 paper, Boltzmann realised that the key lay with the above distribution function and in 1871 he was able to demonstrate the existence of an entropy function in purely mechanical terms and lay down a procedure for finding it.

The following year he took the next step and analysed the behaviour of his entropy function in irreversible processes by considering the way in which the velocity distribution changed with time due to intermolecular collisions. This work can be understood as both an attempt to complete the statistical mechanical reduction of the Second Law, and as showing that the effect of intermolecular collisions on a gas in a non-equilibrium state would be to drive it to equilibrium as described by Maxwell's law. Boltzmann was able to show that the Maxwellian distribution represents the equilibrium state and obtained an explicit formula for the rate of change of f, $\partial f/\partial t$, on the basis of an "exact consideration of the collision process" between two molecules of a spatially homogenous, low-density gas with no external forces present. With this formula in hand, he was able to show that f always tends towards the Maxwellian form. To do this he introduced an auxiliary quantity E (which later came to be called H), defined by,

$$E = \int_0^\infty f(xt)\left\{\log\left[f(x,t)\right]t / \sqrt{x} - 1\right\} dx$$

By considering the symmetrical character of the collision and the possibility of inverse collisions, and assuming, as Maxwell had before him, that the velocities of the two molecules before they collide are statistically independent, Boltzmann demonstrated that E could only decrease with time; that is,

$$\frac{dE}{dz} \leq 0$$

With H substituted for E, this corresponds to Boltzmann's H-Theorem. E cannot decrease to infinity, of course, and so the distribution function must approach a form for which E has a minimum value and its derivative vanishes. At this value the function has, and can only have, the Maxwellian form.

Thus $-E$ increases in the irreversible approach to equilibrium and hence behaves like the entropy function. Furthermore, $-E$ was actually proportional to the entropy in the equilibrium state. This implies, as Boltzmann himself indicated, an entirely new approach to proving the Second Law, one that could deal with the increase in entropy in irreversible processes as well as with its existence as an equilibrium state function. In this way, the H-Theorem effectively extended the definition of entropy to non-equilibrium situations and completed the kinetic reduction of the Second Law of Thermodynamics.

Now, although Maxwell's distribution function introduced a certain statistical 'coarse-graining' into the description, the development of the *H*-Theorem was still based upon consideration of binary collisions between molecules traversing distinct continuous trajectories. Thus, despite the probabilistic elements, the underlying basis was still, of course, Newtonian and deterministic. The tensions this caused within Boltzmann's edifice are well known.

In particular, his conclusion, that the E/H-function always decreased, was disputed by Maxwell, Tait and Thompson and, independently and famously, by Loschmidt, who noted that in any system, "the entire course of events will be retraced if at some instant the velocities of all its parts are reversed". If entropy is a specifiable function of the positions and velocities of the particles of a system, and if that function increases during some particular motion of the system, then reversing the direction of time in the equations of motion will specify a trajectory through which the entropy must decrease. For every possible motion that leads towards equilibrium, there is another, equally possible, that leads away and is therefore incompatible with the Second Law. Loschmidt concluded that if the kinetic theory were true, then the Second Law could not hold universally and thus Boltzmann's proof of the *H*-Theorem could not be correct.

Boltzmann's 1877 response to the Loschmidt 'paradox' is illuminating. He basically admitted that one could not, in fact, prove that entropy increased 'with absolute necessity' and that, according to probability theory, even the most improbable non-uniform distribution is still not absolutely impossible. However, he then claimed that the existence of such improbable entropy-decreasing situations did not contradict the fact that for the overwhelming majority of initial states the entropy could be counted on to increase and that the improbabilities associated with the former case were, for all practical purposes, impossibilities. Boltzmann now focused on the probabilistic aspect of the *H*-Theorem and argued that it followed from this that the number of states leading to a uniform, equilibrium distribution after a certain time must be much larger than the number leading to a non-uniform distribution, since, he claimed, there are infinitely many more uniform states than non-uniform ones. No justification was actually given for this last claim, although Boltzmann intimated where one might be found:

One could even calculate the possibilities of the various states from the ratios of the number of ways in which their distributions could be achieved, which could perhaps lead to an interesting method of calculating thermal equilibrium.

This 'interesting method' was subsequently elaborated later that same year in one of the most significant papers of classical statistical mechanics, entitled 'Probabilistic Foundations of Heat Theory'.

Prior to this Boltzmann had, in a series of papers from 1868 and 1871, re-derived and extended Maxwell's results and, significantly, in 1868, sketched an alternative derivation that was free from any assumptions regarding inter-molecular collisions. Thus he considered the distribution of a fixed amount of energy over a finite number of molecules in such a way that all combinations were equally probable. By regarding the energy as divided into small but finite packets, Boltzmann could treat this as a problem in combinatorial analysis. In this manner he obtained a complicated expression which reduced to Maxwell's law in the limit of an infinite number of molecules and infinitesimal energy elements. This marks the beginnings of his Combinatorial Approach. In the 1877 work, Boltzmann drew on this earlier work and presented a new and radical alternative to the *H*-Theorem approach. In line with his general philosophical attitude towards physical theories this 'Combinatorial Approach' was elaborated through a succession of models of increasing complexity and closer approximation to the physical situation.

The first such model was a highly simplified and explicitly fictional discrete energy model in which he considered a collection of molecules whose individual energies were restricted to a finite, discrete set, with the total energy held fixed. If ω_k is the number of such molecules with energy $k\varepsilon$, then the set $\omega_0,^{\alpha}$ $1.,^{\varepsilon}$ p is sufficient to define a particular macro-state (Zustandverteilung) of the gas. Boltzmann then noted that such a macro-state could be achieved in many different ways, each of which he called a 'complexion'. In general, if a complexion was specified by a set of numbers, each fixing the energy $k\varepsilon$ of the *i*th molecule, then, he wrote, a second complexion belonging to the same macro-state would be achieved by any permutation of the two molecules *i* and *j* which have different energies. Thus a permutation of particles between different energy states gave rise to a different macro-state of the system as a whole.

MECHANICS: MATTER AND ENERGY

Quantum Mechanics is the study of matter and energy on the subatomic scale. It was came into being because classical physics could not explain certain experimental results, such as why the photoelectric effect occured only when the light shone is above a certain frequency.

Principle of Quantatization

The fundamental principle of Quantum Mechanics, and its namesake, is that energy came in discrete particles called "quanta". This idea, the Quantatization of Energy, was put forth by Planck in his Quantum Hypothesis(1900), an explanation of blackbody radiation. It was a departure from Classical Physics, which asserted that energy could be infinately small.

Photoelectric Effect: This quantatization of energy approach was used by Einstein in his 1905 explanation of the photoelectric effect, the emission of electrons by a metal when light is shone on it. No electrons are emitted when the light is below a threshold frequency, no matter how intense the light. Einstein postulated that light, while exibiting wave properties as shone by the Double-slit Experiment, also came in the from of particles called "photons", discrete packets of energy.

The energy of a photon is proportional to the frequency of the light(E=hv); thus, light with higher frequency had more energy and could excite the electrons, while light whose frequency were below the threshold could not, no matter how many photons there were.

Bohr's Model and Spectral Lines

Bohr also used the Quantatization of energy in his 1913 model of the atom. He stated that electrons revolved around the nucleus much like the planets revolve around the sun.

They could be found in predefined, stationary orbits, radiating no energy as they revolve. Only when they gain a quantum of energy equivalent to the difference between the energy of the orbits do they become excited and enter a higher energy orbit.

This is consistent with Einstein's explanation of the photoelectric effect, and explains why spectral lines only occur at certain places.

Wave-Particle Duality of Matter

In 1923 De Broglie expanded the Principle of Wave-particle Duality of Light put forth by Einstein to include all matter, called Principle of Wave-particle Duality of Matter. He asserted that all matter, not just light, exibit both wave-like and particle-like behaviour,

Quantum Matrix and Wave Mechanics

By now there was enough theories to put together a quantum picture of the atom. However, it was very difficult to sort through all the math. Heisenherg and Shroedinger simultaneously and independently worked to put forth a complete quantum mechanics. Heisenberg employed use of matrices, and his model was called the matrix mechanics.

His approach emphasized the quantum—discrete properties while Schroedinger concentrated on the wave properties. He tried to find a function that would encompass all informatin about a particle or system at any given time. This is called the wave function. While debated ensued about which was correct, Schoedinger proved shortly after that the matrix mechanics and the wave function were mathematically equivalent.

Uncertainty Principle

Heisenberg then came up with the Uncertainty Principle, which stated that the position and velocity(momentum, which is mass*velocity) cannot be known precisely simultaneouely. The more accurately one is known, the less accurately the other can be measured. The product of the uncertainty in position, and the uncertainty in momentum, is always greater than or equal to h/(2pi). While this is neglegible on the macroscopic level, subatomically, it causes great uncertainty. As a result, the world exists in statistical probabilities. For example, a electron cannot be found precisely at one location. At any given chance, it has a certain probability of being found anywhere in the universe, with higher probabilities in regions around the atomic nucleus and less likelihood elsewhere. This can be calculated with the wave function.

Quantum Atomic Model

The quantum model of the atom calls for arbitrarily defined orbitals. An orbital is a region around the nucleus where an electron has 90% chance of being found. Electrons have spin, a term used for the intrinsic angular momenta subatomic particles have. They have either a spin of +1/2 or –1/2. Two electrons in the same orbital of the same energy level must have opposite spin (Pauli Exclusion Principle).

Schroedinger's Cat

Since the wave function describes the world in terms of statistical probabilities and not certainties, a cat placed in an isolated system with equal chances of being found alive and dead at anytime is equally dead and alive at the same time. When an observer is introduced to the isolated system, the wave function collapses and the cat is either dead or alive.

Nucleus

The nucleus is a compact and dense region in the centre of the atom that makes up for most of its mass. It is made up of a group of subatomic particles known collectively as nucleons. The two states of nucleons are protons and neutrons.

Although the electron, and not the nucleus, is responsable for chemical reactions, the nucleus still holds much information about the atom. Since nucleons are about 2000 times heavier than electrons, the mass of the nucleus is approximately equal to the mass of the atom.

The number of protons in the nucleus determines the type of atom, and nuclear charge is the property by which the elements are arranged in the periodic table.

Nuclear Energy

A tremendous amount of energy is used to hold the nucleus together, since the repulsion between the positively charged protons is great. The bonding forces that hold the nucleus together are collectively known as the nucleur forces. When heavy element nuclei are split into smaller nuclei, or light element nuclei combined to make one nucleus, tremendous energies are released. The former is known fission, and the latter, fusion, and they are the driving forces behind stars and nuclear weopons. From Dalton's theory, it was thought that atoms were indivisible particles that made up everything in the world. When Rutherford made his gold foil experiment, however, he found that when high-velocity positive alpha particles bombarded a thing metal foil, while most passed straight through, some were reflected back or deflected. This led him to postulate that most of the atom was empty space, with most of the mass concentrated in one dense region in the centre of the arom.

Thus, most of the positive alpha particles passed through the empty space, but some hit the dense core and were reflected back by electric repulsion. He called this the nucleus of the atom.

Subsequently, improvements were made on the Rutherford model. The nucleu was theorized to contain both positive and neutral particles. The interactions in the nucleus, the strong and the weak forces, are the strongest forces in nature. Processes such as fission and fusion release much of this energy, and make possible new energy sources.

In fact, fusion is the process by which stars generate light and heat, and thus the source from which all life on earth is derived. Other processes, such as radiation, have medical applications.

FUNDAMENTALS OF STATISTICS

By statistics we denote the investigation of regularities in apparently non- deterministic processes. An important basic quantity in this context is the "relative frequency" of an "event". Let us consider a repeatable experiment - say, the throwing of a die - which in each instance leads to one of several possible results e - say, $e \equiv$ number of Points = 4. Now repeat this experiment n times under equal conditions and register the number of cases in which the specific result e occurs; call this number $f_n(e)$. The relative frequency of e is then defined as $r(e) \equiv f_n(e)/n$. Following R. von Mises we denote as the "probability " of an event e the expected value of the relative frequency in the limit of infinitely many experiments:

$$p(e) \equiv \lim_{n \to \infty} \frac{f_n(e)}{n}$$

Example: Game die; 100-1000 trials; $e \equiv$ {no. of po int $s = 6$} , or,

$e \equiv$ {no. of po int s ≤ 3} .

Now, this definition does not seem very helpful. It implies that we have already done some experiments to determine the relative frequency, and it tells us no more than that we should expect more or less the same relative frequencies when we go on repeating the trials. What we want, however, is a recipe for the prediction of $p(e)$. To obtain such a recipe we have to reduce the event e to so-called "elementary events" $\in$ that obey the postulate of equal a priori probability. Since the probability of any particular one among K possible elementary events is just $p(\in_i) = 1/K$, we may then derive the probability of a compound event by applying the rules,

$$p(e = \in_i \text{ or } \in_j) = p(\in_i) + p(\in_j)$$

$$p(e = \in_i \text{ and } \in_j) = p(\in_i) \cdot p(\in_j)$$

Thus the predictive calculation of probabilities reduces to the counting of the possible elementary events that make up the event in question.

Example: The result $\in\sigma \equiv$ {no. of po int s = 6} is one among 6 mutually exclusive elementary events with equal a priory probabilities (= 1/6). The compound event $e \equiv$ {no. of po int s ≤ 3} consists of the elementary events $\in_i$ {1, 2 or 3}; its probability is thus p(IV 2V3) = 1/6 + 1/6 = 1/2.

How might this apply to statistical mechanics? - Let us assume that we have N equivalent mechanical systems with possible states $s_1, s_2 \ldots s_K$. A relevant question is then: what is the probability of a situation in which,

$e \equiv$ {k_1 systems are in state s_1, k_2 Stems in State s_2, etc.}

Example: N = 60 dice are thrown (or *one* die 60 times!). What is the probability that 10 dice each have numbers of points 1, 2...6 ? What, in contrast, is the probability that *all* dice show a "one"? The same example, but with more obviously physical content: Let N = 60 gas atoms be contained in a volume V, which we imagine to be divided into 6 equal partial volumes. What is the probability that at any given time we find $k_i = 10$ particles in each subvolume? And how probable is the particle distribution (60, 0, 0, 0, 0, 0)?

We can generally assume that both the number N of systems and the number K of accessible states are very large - in the so-called "thermodynamic limit" they are actually taken to approach infinity. This gives rise to certain mathematical simplifications. Before advancing into the field of physical applications we will review the fundamental concepts and truths of statistics and probability theory, focusing on events that take place in number space, either *R* (real numbers) or N (natural numbers).

Distribution Function

Let x be a real random variate in the region (a, b). The distribution function,

$$p(x_0) \equiv p(x < x_0)$$

is defined as the probability that some x is smaller than the given value x_0. The function $p(x)$ is monotonically increasing and has $P(a) = 0$ and $p(b) = 1$. The distribution function is dimensionless: $[P(x) = 1]$.

The most simple example is the equidistribution for which,

$$P(x_0) = \frac{x_0 - a}{b - a}$$

Another important example, with $a = -\infty$, $b = \infty$, is the normal distribution,

$$p(x_0) = \frac{1}{\sqrt{2\pi}} \int_{-\infty}^{x0} dxe^{-x2/2}$$

and its generalisation, the Gaussian distribution,

$$p(x_0) = \frac{1}{\sqrt{2\pi\sigma^2}} \int_{-\infty}^{x0} dxe^{-(x-(x))^2/2\sigma^2}$$

where the parameters (x) and σ^2 define an ensemble of functions.

Distribution Density

The distribution or probability density p(x) is defined by $p(x_0)\, dx \equiv p\{x \in [x_0 + dx]\} \equiv dP\,(x_0)$

In other words, $p(x)$ is just the differential quotient of the distribution function:

$$p(x) = \frac{dP(x)}{dx}, i.e. P(x_0) = \int_a^{x0} p(x)dx$$

$p(x)$ has a dimension; it is the reciprocal of the dimension of its argument,

$$[p(x)] = \frac{1}{[x]}:$$

For the equidistribution we have,

$$p(x) = 1/(b-a),$$

and for the normal distribution,

$$p(x) = \frac{1}{\sqrt{2\pi}} e - x^2 / 2$$

If x is limited to discrete values x_a with a step $\Delta x_a \equiv x_a + 1\ x_a$ one often writes,

$$p_a \equiv p(x_a)\ \Delta x_a$$

for the probability of the event x_a. This $\Delta x_a \equiv x_{a+1} - x_a$ is by definition dimensionless, although it is related to the distribution density $p(x)$ for continuous arguments. The special case that x is restricted to integer values k; in that case $\Delta x_a = 1$.

Moments of a Density

By this we denote the quantities,

$$\langle x^n \rangle \equiv \int_a^b x^n p(x) dx \text{ or in the Discrete case, } \equiv \sum_a p_a x_a^n$$

The first moment $\langle x \rangle$ is also called the expectation value or mean value of the distribution density $p(x)$, and the second moment (x^2) is related to the variance and the standard deviation: variance $\sigma^2 = \langle x^2 \rangle - \langle x \rangle^2$ (standard deviation σ = square root of variance).

Some Important Distributions

- **Equidistribution:** Its great significance stems from the fact that this distribution is central both to statistical mechanics and to practical numerics. In the theory of statistical-mechanical systems, one of the fundamental assumptions is that all states of a system that have the same energy are equally probable (axiom of equal a priori probability). And in numerical computing the generation of homogeneously distributed pseudo-random numbers is relatively easy; to obtain differently distributed random variates one usually "processes" such primary equidistributed numbers.

- **Gauss distribution:** This distribution pops up everywhere in the quantifying sciences. The reason for its ubiquity is the "central value theorem": Every random variate that can be expressed as a sum of arbitrarily distributed random variates will in the limit of many summation terms be Gauss distributed. For example, when we have a complex measuring procedure in which a number of individual errors (or uncertainties) add up to a total error, then this error will be nearly Gauss distributed, regardless of how the individual contributions may be distributed. In addition, several other physically relevant distributions, such as the binomial and multinomial densities, approach the Gauss distribution under certain - quite common -circumstances.

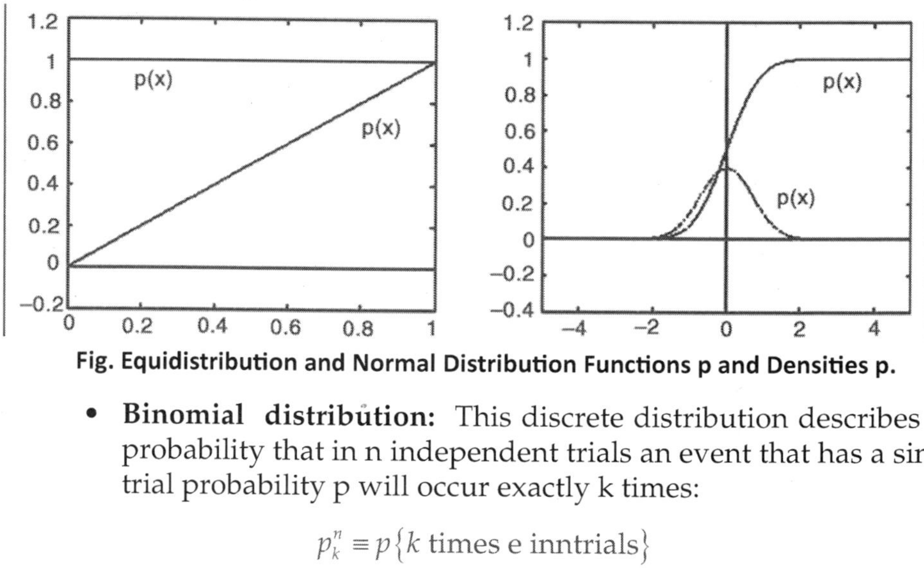

Fig. Equidistribution and Normal Distribution Functions p and Densities p.

- **Binomial distribution:** This discrete distribution describes the probability that in n independent trials an event that has a single trial probability p will occur exactly k times:

$$p_k^n \equiv p\{k \text{ times e inntrials}\}$$

$$= \binom{n}{k} p^k (1-p)^{n-k}$$

For the first two moments of the binomial distribution we have $\langle k \rangle = np$ (not necessarily integer) and $\sigma^2 = np\,(1-p)$ (i.e. $\langle k^2 \rangle - \langle k2 \rangle^2$).

Application: Fluctuation processes in statistical systems are often described in terms of the binomial distribution. For example, consider a particle freely roaming a volume V. The probability to find it at some given time in a certain partial volume V_1 is $p(V_1) = V_1/V$. Considering now N independent particles in V, the probability of finding just N_1 of them in V_1 is given by,

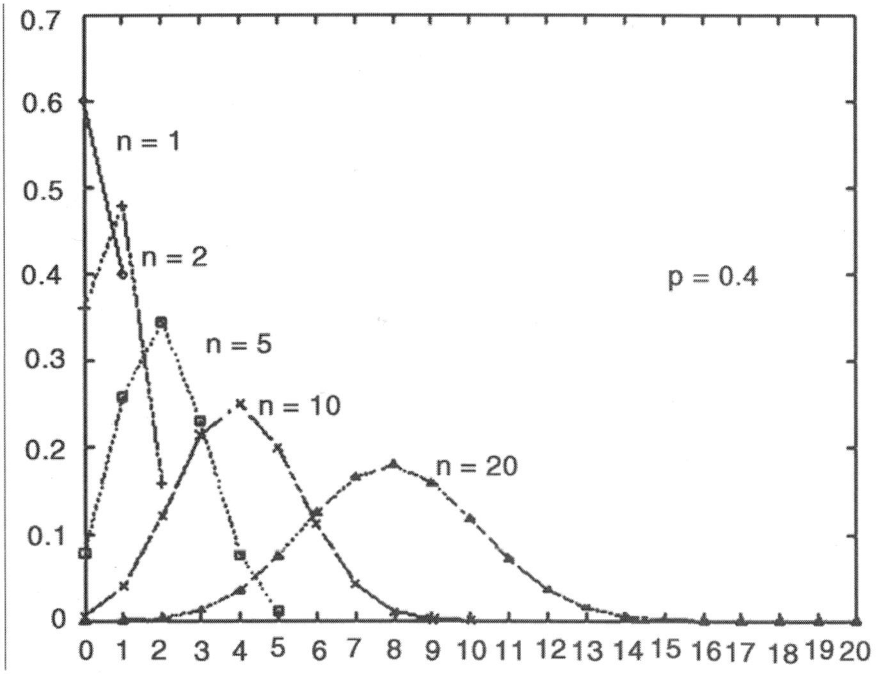

Fig: Binomial Distribution Density p_k^n .

$$p_{N_1}^N = \binom{N}{N_1}(V_1/V)^{N_1}(1 - V_1/V)N - N_1$$

The average number of particles in V_1 and its standard deviation are,

$$\langle N_1 \rangle = NV_1/V \text{ and } \sigma \equiv \sqrt{(\Delta N_1)2} = \sqrt{N(V_1/V)(1 - V_1/V)}$$

Note that for Np ≈ 1 we have for the variance $\sigma^2 \approx \sigma \approx \langle N_1 \rangle \approx 1$, meaning that the population fluctuations in V_1 are then of the same order of magnitude (namely, 1) as the mean number of particles itself. For large n such that $np >> 1$ the binomial distribution approaches a Gauss distribution with mean np and variance σ^2 = npg (theorem of Moivre-Laplace):

$$p_k^n \Rightarrow pg(k) = \frac{1}{\sqrt{2\pi npq}} \exp\left[-(k - np)^2 / 2npq\right]$$

with $q = 1 - p$.

If $n \to \infty$ and $p \to 0$ such that their product $np \equiv \lambda$ remains finite,

$$P_n(k) = \frac{\lambda^k}{K!} e^{-\lambda}$$

which goes by the name of Poison distribution.

An important element in the success story of statistical mechanics is the fact that with increasing n the sharpness of the distribution 1.39 or 1.42 becomes very large. The relative width of the maximum, i.e. $\sigma / (k)$, decreases as $1/\sqrt{n}$. For $n = 10^4$ the width of the peak is no more than 1 per cent of (K), and for "molar" orders of particle numbers $n \approx 10^{24}$ the relative width $\sigma / \langle k \rangle$ is already $\approx 10^{-12}$. Thus the density approaches a "delta distribution". This, however, renders the calculation of averages particularly simple:

$$(f(k)) = \sum_k p_n^k f(k) \rightarrow f(\langle k \rangle)$$

or,

$$\langle f(k) \rangle \approx \int dk \delta(k - (k) f(k))$$

- **Multinomial distribution:** This is a generalisation of the binomial distribution to more than 2 possible results of a single trial. Let $e_1, e_2 ..., e_k$ be the (mutually exclusive) possible results of an experiment; their probabilities in a single trial are $p_1, p_2 ... p_k$, with $\sum_i^K P_1 = 1$. Now do the experiment n times; then,

$$p_n(k_1, k_2 ..., \kappa_K) = \frac{n!}{\kappa_1! \kappa_2! ... \kappa_K!} p_1^{\kappa_1} p_{p_2}^{\kappa_2} ... p_k^{\kappa k} \left(\text{with } \kappa_1 + \kappa_2 + ... + \kappa k = n\right)$$

is the probability to have the event e_1 just k_1 times, e_2 accordingly k_2 times, etc.

We get an idea of the significance of this distribution in statistical physics if we interpret the K possible events as "states" that may be taken on by the n particles of a system (or, in another context, by the n systems in an ensemble of many-particle systems). The above formula then tells us the probability to find k_1 among the n particles in state e_1, etc. Example: A die is cast 60 times. The probability to find each number of points just 10 times is,

$$p\infty(10,10,10,10,10) = \frac{60!}{(10!)^{\sigma}} \left(\frac{1}{6}\right)^{\infty} = 7.457.10^{-5}$$

To compare, the probabilities of two other cases: $p\infty(10,10,10,9,11)$ = $6.778.10^{-5}$, $p\infty(15,15,15,5,5,5)$ = $4.406.10^{-s}$. Finally, for the quite improbable case (60,0,0,0,0,0) we have $p\infty = 2.046.10^{-47}$

Due to its large number of variables (k) we cannot give a graph of the multinomial distribution. However, it is easy to derive the following two important properties: Approach to a multivariate Gauss distribution: just as the binomial distribution approaches, for large n, a Gauss distribution, the multinomial density approaches an appropriately generalised - "multivariate" - Gauss distribution. Increasing sharpness: if n and $k_1...k_k$ become very large (multiparticle systems; or ensembles of $n \to \infty$ elements), the function p_n (k_1, k_2..., $\kappa k \equiv p_n\left(\vec{k}\right)$ has an extremely sharp maximum for a certain partitioning $\vec{k}^* \equiv \left\{k_1^*, k_2^*, ..., \kappa_K^*\right\}$, namely $\left\{k_1^* - npi;_i, ..., K\right\}$. This particular partitioning of the particles to the various possible states is then "almost always" realised, and all other allotments (or distributions) occur very rarely and may safely be neglected. This is the basis of the method of the most probable distribution which is used with great success in several areas of statistical physics.

Stirling's Formula

For large values of m the evaluation of the factorial m! is difficult. A handy approximation is Stirling's formula,

$$m! \approx \sqrt{2\pi m}\,(m / e)^m$$

Example: $m = 69$ (Near most pocket calculators' limit): $69! = 1.7112.10^{98}$; $\sqrt{2\pi.69}\,(69 / e)^{\sigma 9} = 1.7092.10^{98}$.

The same name Stirling's formula is often used for the logarithm of the factorial:

$$\text{In m!} \gg \text{m (In m} - 1) + \text{In} \sqrt{2\text{ pm}} \gg \text{m (In m} - 1)$$

(The term $In\ \sqrt{2\pi m}$ may usually be neglected in comparison to m (Inm – 1).

STATISTICAL MECHANICS FUNDAMENTALS

Most traditional presentations of statistical mechanics begin by expressing entropy in terms of W, the number of equally probable microscopic states that constitute a given macroscopic state. For example, a "macroscopic" state where 4 pupils (molecules) are seated in any 6 chairs can be achieved in W = 30 "microscopic" ways. Here we develop the relevant form of entropy using the dynamic interpretation of temperature.

Boltzmann Entropy

Let W represent the total number of microstates. From the dynamic interpretation of temperature, the ratio dQ/kT is the number of new states available to the system due to the introduction of heat dQ. It follows that the change in total microstates dW is given by,

$$dW = W\frac{dQ}{kT}$$

Introduce entropy using $dS = dQ/T$ to obtain,

$$kdW = WdS.$$

Solving, we find the Boltzmann form for entropy,

$$S = k\ In\ W$$

- (a) Evaluate microstates W_1 for one particle of a monatomic ideal gas in volume V using the (crude) idea that each particle occupies a cube of volume λ^3 were the sides are one deBroglie wavelength λ. The number of ways one particle can fit in V is V/λ^3 and the average energy per particle is $U/N=3kT/2$. (Recall the deBroglie relation $\lambda = h/mv$.) (b) Form the total number of microstates for N particles, $W = W_1^N$ and calculate the entropy S and use it to find an expression for the adiabatic expansion of an ideal monatomic gas, Equation. (Use S = constant or $S_1 = S_2$) For now, we assume the particles are distinguishable from each other.
- (a) Repeat the development of problem 6 for radiation (photon gas).
  Show that an adiabatic condition for radiation is $T_1T_1^3 = V_2T_2^3$.
  (b) The background radiation from the Big Bang is now 2.7 K. What was the temperature when the universe was 1.0×10^{-10} of its current volume?

Entropy and Thermodynamics

Statistical mechanics is likely to provide us with an entropy representation. Then (2) can be rewritten to express dS, Once a system's entropy is calculated from statistical mechanics, all of the related thermodynamic properties can be generated. Here is the principle trick that lets us wring thermodynamic state equations from entropy.

$$dS = \frac{dU}{T} + \frac{PdV}{T}$$

This can be compared with the mathematical statement,

$$dS = \frac{\partial S}{\partial U}dU + \frac{\partial S}{\partial V}dV$$

math to obtain relations between S and the more accessible variables T and P:

$$\frac{1}{T} = \frac{\partial S}{\partial U} \text{ and } \frac{P}{T} = \frac{\partial S}{\partial V}$$

- In problem 6, you found that entropy for a monatomic ideal gas is given by $S = Nk \text{ In} \left[\frac{V}{h^3} \left(\frac{2mU}{N} \right)^{\frac{3}{2}} \right]$. Use the relations (11) to derive the ideal gas law and the energy relation for monatomic gas. Show that the results give the ideal gas equation and the relation $U = 3/2\, nR\, T$.
- An entropy representation for radiation from heated objects is,

$$S = \frac{4}{3}\left(\sigma V U^3\right)^{\frac{1}{4}}$$

where s is a constant. Find the state equations for this "blackbody radiation." Express U entirely in terms of T and V and eliminate T to express P entirely in terms of U and V, and (these are the most familiar forms).

Canonical Distribution

Here we derive the probability that a system with energy levels E_1, E_2... at temperature T is in any particular energy level E_i. The probabilities p_j are used to find averages of physical quantities. For instance, the average energy E of a particle with possible energies E_j available is given by,

$$E = \sum_i p_i E_i$$

- A two-state system has energy levels of 1.0 eV and 2.0 eV. The probability that the system is in the lower energy state is 0.75. Find the average energy of the system.

Introduce the Helmholtz free energy F as the minimum work required to assemble the system from its well-separated components at constant temperature. Denote the average energy exchanged per interaction as kT. We want to evaluate the probability p_j that the system is in level j. The energy available for changing the state from level j is $E_j - F$. Consider an arbitrary time interval during which there are N interactions. The energy available for transitions is then NkT and the probability of a transition out of state j is $(E_j - F)/N\,kT$.

Each interaction then has a complementary probability $1-(E_j - F)/NkT$ of not causing the system to leave level j. The probability that none of the N interactions causes a transition is p_j,

$$p_j = \left(1 - \frac{E_j - F}{NkT}\right)^N$$

In the limit of large N this becomes1 the sought-for level probability,

$$p_j = \exp\left(\frac{F - E_j}{kT}\right)$$

Notice that a lower energy level Ei is favoured over a higher level. Also note that the higher the temperature, the closer all exponents approach to zero, so that the levels become more equally populated.

- The ground state and first excited states of hydrogen atoms are respectively –13.6 eV and –3.4 eV. Find the ratio of ground state atoms to first excited state atoms at (a) 3000 K and (b) 6000 K (the temperature at the surface of the sun).

Application: Paramagnetism

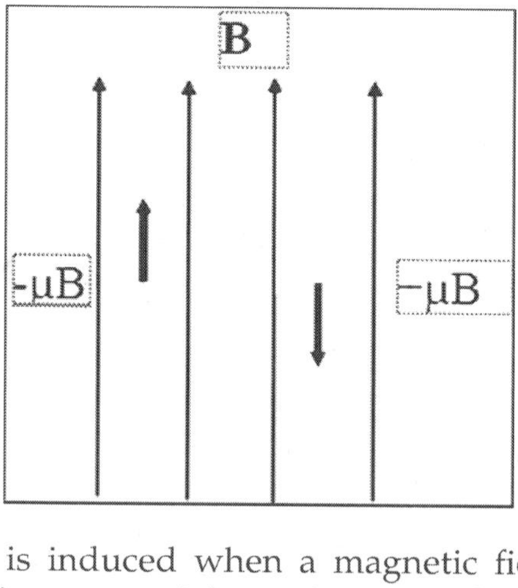

Paramagnetism is induced when a magnetic field B interacts with unpaired electrons in a material causing a preferential orientation of magnetic moments. A simplified model is a uniform magnetic field applied to a collection of free electrons. Quantum mechanics dictates that only two states are allowed, parallel and antiparallel. When the B field and the spin orientation are parallel, the energy per electron is $E_{\uparrow} = -\mu B$ where m is the magnetic moment of the electron. Conversely, an antiparallel orientation has energy $E_{\downarrow} = +\mu B$.

(a) Use the canonical distribution to write the respective probabilities $p_\uparrow$ and $p_\downarrow$ that a single electron is oriented parallel or antiparallel to a uniform B-field. (We will soon evaluate the factor exp(F/kT), but it will cancel out in this problem)

(b) Given that there are N total electrons in the material, write expressions for the numbers of electrons $N_\uparrow$ and $N_\downarrow$ oriented parallel and antiparallel to the field.

(c) Show that the total energy $\overline{E} = N_\uparrow E_\uparrow + N_\downarrow E_\downarrow$ is given by $\overline{E} = -N\mu B\left[\tanh\left(\frac{\mu B}{kT}\right)\right]$

PARTITION FUNCTION

Write the normalisation condition,

$$Z = \Sigma \exp\left(\frac{-E_i}{kT}\right)$$

and substitute from Equation (above) to obtain a fundamental connection with thermodynamics,

$$F = -kT \ln Z$$

where the partition function Z is defined as,

$$Z = \Sigma \exp\left(\frac{-E_i}{kT}\right)$$

Moreover, Equation (above) can be rewritten using Equation (above) to put the canonical probabilities in a more traditional form,

$$p_j = \frac{1}{Z}\exp\left(\frac{-E_j}{kT}\right)$$

- Use equations above to derive equations above.
- From memory, (a) write the canonical distribution probability and identify the symbols in the expression (b)Write an expression for the partition function Z and (c) express Helmholtz free energy F in terms of Z. (In later chapters, we will wring thermodynamic information from both F and Z.)
- A particular enzyme has two states, (i) inactive, with energy E, and (ii) activated, with energy E + 0.003 aJ. (a = atto = $\times 10^{-18}$).
(a) Write the probability of each state at 310 K. (b) Find the sum of the probabilities. (c) Which state occurs with greater frequency?

Looking Ahead

A few things should be mentioned as we look ahead to further elaborations and consolidations. Entropy is a cornerstone for statistical theory and we focused on entropy as the link between microphysics and macrophysics. However, we will find that other thermodynamic quantities like *U*, *F*, and *Z* and others often serve as the link even more simply or directly.

This overview presented two powerful concepts that connect the micro-world to the macro-world, entropy and probability. The next few chapters develop a most useful connection between these two. We will reconstruct entropy in terms of probabilities to find the Gibbs or Shannon form $S = -kN \Sigma p_i ln\ p_i$

- Our plan for this course is to first introduce information theory and derive the canonical distribution and other probabilities based on the maxent principle. We will also develop other distributions that are useful for isolated systems and for systems that exchange particles with the environment (the micro canonical and grand canonical distributions). Basics of classical thermodynamics will be developed as byproducts. We will present shortcuts to the averaging processes based on the partition function. Most importantly, we will use our results for applications to various branches of physics and chemistry.

MATHEMATICAL STATISTICS

Mathematical statistics uses probability theory and other branches of mathematics to study statistics from a purely mathematical standpoint.

Mathematical statistics deals with gaining information from data. In practice, data often contain some randomness or uncertainty. Statistics handles such data using methods of probability theory.

Statistics is divided into:

- *Descriptive statistics:* the part of mathematical statistics that describes data, i.e. summarises the data and their typical properties
- *Inferential statistics:* the part of mathematical statistics that draws conclusions from data, i.e. checks whether the data fulfill some condition and gives guarantees on the involved uncertainty.

Mathematical statistics is the theoretical basis for many practices in applied statistics.

DESCRIPTIVE STATISTICS

Descriptive Statistics are used to describe the basic features of the data in a study. They provide simple summaries about the sample and the measures. Together with simple graphics analysis, they form the basis of virtually every quantitative analysis of data. Various techniques that are commonly used are classified as:

1. Graphical description in which we use graphs to summarize data.
2. Tabular description in which we use tables to summarize data.
3. Summary statistics in which we calculate certain values to summarize data.

In general, statistical data can be described as a list of subjects or units and the data associated with each of them. Although most research uses many data types for each unit, we will limit ourselves to just one data item each for this simple introduction.

We have two objectives for our summary:

1. We want to choose a statistic that shows how different units seem similar. Statistical textbooks call the solution to this objective, a *measure of central tendency.*
2. We want to choose another statistic that shows how they differ. This kind of statistic is often called a *measure of statistical variability.*

When we are summarizing a quantity like length or weight or age, it is common to answer the first question with the arithmetic mean, the median, or the mode. Sometimes, we choose specific values from the cumulative distribution function called quantiles.

The most common measures of variability for quantitative data are the variance; its square root, the standard deviation; the range; interquartile range; and the average absolute deviation (average deviation).

One important use of descriptive statistics is to summarize a collection of data in a clear and understandable way. For example, assume a psychologist gave a personality test measuring shyness to all 2500 students attending a small college. How might these measurements be summarized? There are two basic methods: numerical and graphical. Using the numerical approach one might compute statistics such as the mean and standard deviation. These statistics convey information about the average degree of shyness and the degree to which people differ in shyness. Using the graphical approach one might create a stem and leaf display and a box plot. These plots contain detailed information about the distribution of shyness scores.

Graphical methods are better suited than numerical methods for identifying patterns in the data. Numerical approaches are more precise and objective. Since the numerical and graphical approaches compliment each other, it is wise to use both.

STATISTICAL INFERENCE

Inferential statistics or statistical induction comprises the use of statistics to make inferences concerning some unknown aspect of a population. It is distinguished from descriptive statistics. Statistical inference concerns the problem of inferring properties of an unknown distribution from data generated by that distribution. The most common type of inference involves approximating the unknown distribution by choosing a distribution from a restricted family of distributions. Generally the restricted family of distributions is specified parametrically.

POWER LAW DISTRIBUTION IN STATISTICAL MECHANICS

From the notion of statistical mechanics, it is known that distributions dominate the nature phenomenon and the world structures. Some of them are discovered and well interpreted; some of them are invisible to us or beyond our understanding. Barabási and Albert in 1999 suggest a power law distribution which was found in a number of complex networks. They are called scale free networks. Such kind of power law distribution can describe a variety of systems both in the nature, society and man-made high-tech visual world, such as the World Wide Web, which is an enormous visual network of human intelligence. In this network, the nodes are the documents and the edges are the hyperlinks that point from one document to another. It is discovered that the degree distribution of web pages follows a power law distribution. Both probability that a document has k outgoing hyperlinks, Pour, (k) and the distribution of incoming edges, pin (k) have power-law tails.

This can be interpreted as the more links one has today, the more links it will have tomorrow. However, this is only on the surface level of the experiment understanding. The deep physics meaning are not revealed.

We try to understand such phenomenon from different perspective by treating networks as thermodynamics system. At first glance, it sounds rude because network is not highly thermalised. However, further reflection tells that it is actually its intrinsic properties that we have ignored before. Take social networks for example.

As members of a community and citizens of a country, we never stop doing our effort to keep our society stable against the "second law of thermodynamics". Otherwise, the social network will break up. This gives us a hint that, in most cases, the "uncertainty" of a stable network will not change over time.

To interpret it in physics is that the mean entropy of a network is a constant. On this assumption, we will in the following sections show how it works to demonstrate the power law distribution.

INFORMATION THEORY AND MAXIMUM-ENTROPY ESTIMATES

Information theory provides a constructive criterion for setting up probability distributions on the basis of partial knowledge, and leads to a type of statistical inference which is called the maximum entropy estimate. Where the Shannon entropy are usually used:

$$H[P(x)] = -k \Sigma P_i \ln P_i (x)$$

It is the least biased estimate possible on the give information. Suppose,

$$<f(x)> = S \text{ Pi f (xi)}$$

The quantity x is capable of assuming the discrete values x_i $(i = 1, 2\ldots, n)$. We are not given the corresponding probabilities pi; all we know is the expectation value of function $f(x)$; and the normalisation condition.

From the merely facts equations above, we want to find the probability assignment which avoids bias, while agreeing with whatever information is given. Therefore, in making inferences on the basis of partial information we must use that probability distribution which has maximum entropy subject whatever is know.

The Derivation of Power Law Distribution

We consider the mean entropy of system as a constant. In the language of information theory, it means the expectation value of entropy $S(x)$ is known, namely,

$$\langle S(x)\rangle = \sum_i P_i[\Omega_i(x)]S_i(x)$$

P_i is the probability of finding a system in one of the states corresponding to the ith element (or group). where the entropy of each element in the system is represented by the Boltzmann entropy,

$$S_i(x) = k_B \ln[\Omega_i(x)]$$

where the Ω_I is number of states of the ith element (or group).

Application on Networks

Networks are likely to be interpreted as graphs with nodes linked by edges. Simple behaviour of each node (identical or not) can give a complex collective behaviours in a network. However, there are some important quantities that would generalise the topology of a network.

The first one is the degree (the number of edges of a node) distribution. It describes how links distributed among different nodes. Here, we can treat the number of state of the ith element (Ω_i) as the degree of the ith node, because the more connections one node have the more opportunity it can interact the other nodes, namely,

$$\Omega_i \sim n_i$$

where n_i is the degree of the ith node. Then substitute it into equation:

$$Pi = \frac{n_i^{-\alpha}}{\sum_i n_i^{\alpha}} = An_i^{-\alpha}$$

This give us very clear physics meaning: if a network is dynamically stable (or temporarily say stable), the degree distribution would follow the power law. Remember, dynamically stable is very important here. If new nodes are inserted randomly into the network continuously, the power law will break up gradually.

The second one is the aggregation. The aggregation in networks is termed as clustering. Usually, the clusters in a network can be treated as sub-networks, which have various amounts of nodes and are relatively isolated to the other notes of their supplementary network. Take the telecommunication network for example. Local telecom network, such as a network of city, can be treated as a subsystem of the state network. Traffic in such networks are main local calls.

However, the toughest thing to deal with is how we count the number of state of a sub-network, if we want to take advantage. To think about it, we'd better trace back, where we used the opportunity interpretation for a single node. Can we use the same inference? The answer is no but a little bit similar. Suppose a sub-group with M nodes.

The maximum links the network could have is M(M–1)/2, but it is not the number of states of the sub-network because it only relates to the number of nodes in the network and has nothing to do with the other properties of the network.

It seems we have to discover a new quantity to serve our need. However, predecessors of network topology have already solved the problem. They suggest another quantity named average path length which is the average of the shortest distance between any nodes. Therefore the number of states could be easily defined as,

$$\Omega_i = C_{L_i}^{l}$$

Where L is total path length of ith network and l is the average path length; usually in an effective network $L \sim M^{\beta}$

$$0 < \beta \leq 2$$

Then,

$$\Omega_i \cong C^{l}_{M^{\beta}} = \frac{M^{\beta}\left(M^{\beta}-1\right)\cdots\left(M^{\beta}-l+1\right)}{l!} \approx \frac{M^{l\beta}}{l!}$$

where M >> l; substitute it into equation,

$$P_i = BM^{-\beta\alpha}$$

equation (above) shows very clear physics meaning that the size distribution of sub-network follow the power law.

ELEMENTARY STATISTICAL CONCEPTS

In this introduction, we will briefly discuss those elementary statistical concepts that provide the necessary foundations for more specialized expertise in any area of statistical data analysis. The selected topics illustrate the basic assumptions of most statistical methods and/or have been demonstrated in research to be necessary components of one's general understanding of the "quantitative nature" of reality. Because of space limitations, we will focus mostly on the functional aspects of the concepts discussed and the presentation will be very short. Further information on each of those concepts can be found in statistical textbooks.

WHAT ARE VARIABLES

In statistics, variables refer to measurable attributes, as these typically vary over time or between individuals. Variables can be discrete (taking values from a finite or countable set), continuous (having a continuous distribution function), or neither. Temperature is a continuous variable, while the number of legs of an animal is a discrete variable. This concept of a variable is widely used in the natural, medical and social sciences. In causal models, a distinction is made between *"independent variables"* and *"dependent variables"*, the latter being expected to vary in value in response to changes in the former. In other words, an independent variable is presumed to potentially affect a dependent one. In experiments, independent variables include factors that can be altered or chosen by the researcher independent of other factors.

For example, in an experiment to test whether or not the boiling point of water changes with altitude, the altitude is under direct control and is the *independent* variable, and the boiling point is presumed to depend upon it and is therefore the *dependent* variable. The collection of results from an experiment, or information to be used to draw conclusions, is known as data. It is often important to consider which variables to allow for, or to directly control or eliminate, in the design of experiments.

While independent variables can refer to quantities and qualities that are under experimental control, they can also include extraneous factors that influence results in a confusing or undesired manner.

In general, if strongly confounding variables exist that can substantially affect the result, then this makes it more difficult to interpret the results. For example, a study into the incidence of cancer with age will also have to take into account variables such as income (poorer people may have less healthy lives), location (some cancers vary depending on diet and sunlight), stress and lifestyle issues (cancer may be related to these more than age), and so on. Failure to at least consider these factors can lead to grossly inaccurate deductions. For this reason, controlling unwanted variables is important in research. Variables are things that we measure, control, or manipulate in research. They differ in many respects, most notably in the role they are given in our research and in the type of measures that can be applied to them.

A symbol or name that stands for a value. For example, in the expression, $x+y$ in which x and y are variables.

Measurement Scales

Variables differ in "how well" they can be measured, i.e., in how much measurable information their measurement scale can provide. There is obviously some measurement error involved in every measurement, which determines the "amount of information" that we can obtain. Another factor that determines the amount of information that can be provided by a variable is its "type of measurement scale." Specifically variables are classified as (a) nominal, (b) ordinal, (c) interval or (d) ratio.

1. Nominal variables allow for only qualitative classification. That is, they can be measured only in terms of whether the individual items belong to some distinctively different categories, but we cannot quantify or even rank order those categories. For example, all we can say is that 2 individuals are different in terms of variable A (e.g., they are of different race), but we cannot say which one "has more" of the quality represented by the variable. Typical examples of nominal variables are gender, race, colour, city, etc.
2. Ordinal variables allow us to rank order the items we measure in terms of which has less and which has more of the quality represented by the variable, but still they do not allow us to say "how much more." A typical example of an ordinal variable is the socioeconomic status of families. For example, we know that upper- middle is higher than middle but we cannot say that it is, for example, 18% higher. Also this very distinction between nominal, ordinal, and interval scales itself represents a good

example of an ordinal variable. For example, we can say that nominal measurement provides less information than ordinal measurement, but we cannot say "how much less" or how this difference compares to the difference between ordinal and interval scales.

3. Interval variables allow us not only to rank order the items that are measured, but also to quantify and compare the sizes of differences between them. For example, temperature, as measured in degrees Fahrenheit or Celsius, constitutes an interval scale. We can say that a temperature of 40 degrees is higher than a temperature of 30 degrees, and that an increase from 20 to 40 degrees is twice as much as an increase from 30 to 40 degrees.
4. Ratio variables are very similar to interval variables; in addition to all the properties of interval variables, they feature an identifiable absolute zero point, thus they allow for statements such as x is two times more than y. Typical examples of ratio scales are measures of time or space. For example, as the Kelvin temperature scale is a ratio scale, not only can we say that a temperature of 200 degrees is higher than one of 100 degrees, we can correctly state that it is twice as high. Interval scales do not have the ratio property. Most statistical data analysis procedures do not distinguish between the interval and ratio properties of the measurement scales.

Correlational vs. Experimental Research

Most empirical research belongs clearly to one of those two general categories. In correlational research we do not (or at least try not to) influence any variables but only measure them and look for relations (correlations) between some set of variables, such as blood pressure and cholesterol level. In experimental research, we manipulate some variables and then measure the effects of this manipulation on other variables; for example, a researcher might artificially increase blood pressure and then record cholesterol level. Data analysis in experimental research also comes down to calculating "correlations" between variables, specifically, those manipulated and those affected by the manipulation. However, experimental data may potentially provide qualitatively better information: Only experimental data can conclusively demonstrate causal relations between variables. For example, if we found that whenever we change variable A then variable B changes, then we can conclude that "A influences B." Data from correlational research can only be "interpreted" in causal terms based on some theories that we have, but correlational data cannot conclusively prove causality.

Dependent vs. Independent Variables

Independent variables are those that are manipulated whereas dependent variables are only measured or registered. This distinction appears terminologically confusing to many because, as some students say, "all variables depend on something." However, once you get used to this distinction, it becomes indispensable. The terms dependent and independent variable apply mostly to experimental research where some variables are manipulated, and in this sense they are "independent" from the initial reaction patterns, features, intentions, etc. of the subjects. Some other variables are expected to be "dependent" on the manipulation or experimental conditions. That is to say, they depend on "what the subject will do" in response. Somewhat contrary to the nature of this distinction, these terms are also used in studies where we do not literally manipulate independent variables, but only assign subjects to "experimental groups" based on some pre-existing properties of the subjects. For example, if in an experiment, males are compared with females regarding their white cell count (WCC).

RELATIONS BETWEEN VARIABLES

Regardless of their type, two or more variables are related if in a sample of observations, the values of those variables are distributed in a consistent manner. In other words, variables are related if their values systematically correspond to each other for these observations. For example, Gender and WCC would be considered to be related if most males had high WCC and most females low WCC, or vice versa; Height is related to Weight because typically tall individuals are heavier than short ones; IQ is related to the Number of Errors in a test, if people with higher IQ's make fewer errors.

WHY RELATIONS BETWEEN VARIABLES ARE IMPORTANT

Generally speaking, the ultimate goal of every research or scientific analysis is finding relations between variables. The philosophy of science teaches us that there is no other way of representing "meaning" except in terms of relations between some quantities or qualities; either way involves relations between variables. Thus, the advancement of science must always involve finding new relations between variables. Correlational research involves measuring such relations in the most straightforward manner. However, experimental research is not any different in this respect. For example, the above mentioned experiment comparing WCC in males and females can be described as looking for a correlation between two variables: Gender and WCC. Statistics does nothing else but help us evaluate relations between variables.

□

2

Continuum Mechanics

Continuum Mechanics is a means of studying the deformation and flow of matter by ignoring its particulate nature. It is a simplification that makes it possible to investigate the movement of matter on scales larger compared to typical distances between particles. Thus, in studying the movement of air in the atmosphere the fact that is it made up of atoms is ignored. It would, in any case, be impossible to predict the motion of each atom.

Fortunately, when asking questions such as "How warm will it be tomorrow?" we are only interested in the average behaviour of large numbers of atoms, not in their individual motion. Similarly, in the study of mud flow the movement of industrial clay particles is ignored. At larger scales the evolution of galaxies can also be described using continuum models instead of trying to predict the movement of every planet, asteroid, and star. This approach is used to study a huge range of phenomena, including the flow of air and water as well as in studies of rock slides (the individual particles being large rocks), snow avalanches, blood flow, and even galaxy evolution.

The Continuum and Fluid Mechanics courses in the Applied Mathematics department at the University of Waterloo. We hope that the material contained in these pages enriches the experience of those taking our courses (AM 361 and AM 463).

At the same time, for those who are at an earlier point in their university education, we hope to show you both the richness of our field of study as well as illustrate how the courses you take earlier in your career tie in with what we do in our upper year courses and in our research.

Continuum and Fluid Mechanics each have a long history. Indeed we take some past achievements nearly for granted whenever we fly in an airplane or turn on the television to check the weather report.

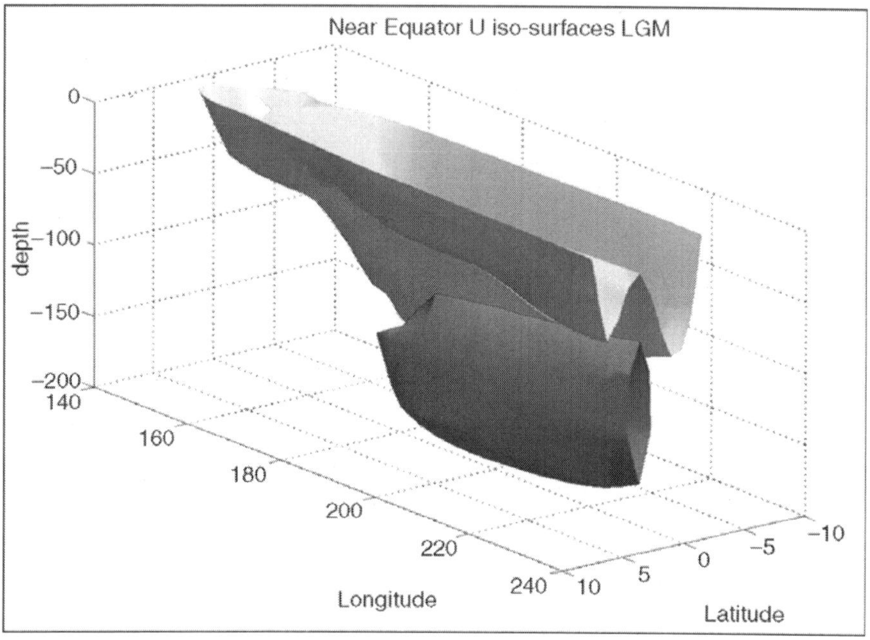

Yet it is likely true that the wide availability of computers for research purposes makes the present a new Golden Age for the study of both fluid and continuum mechanics. It is perhaps easier to define fluid mechanics than continuum mechanics. After all we are surrounded, indeed bathed in, fluids for the entirety of our lives. This is because at the level of their gross, or macroscopic, behaviour (how they flow, for example) both gases and liquids can be considered to be fluids.

However, the actual mathematical description of fluid flow can be vastly different depending on the scales of interest. Thus, while the aerodynamics of a bumble bee's flight and the evolution of a hurricane are technically governed by the same set of equations, the simplifications made to reflect the dominant physics of each situation are wildly different.

As broad as fluid mechanics is, continuum mechanics can be considered to be even broader. This is because it is concerned with the general theory of equations that govern fluids, solids and all the exotic materials in between, from a general, and often abstract, point of view. Moreover, since much of the materials that surround us and make much of modern life possible (plastics being the most obvious example) are neither fluid nor solid, but something in between, there is also an intensely practical side to continuum mechanics. Indeed, the study of materials ranges across many disciplines in both the pure and applied sciences. One of the things we hope to convey throughout these web materials is the particular nature of studying materials within an applied mathematics setting.

CONCEPT OF FLUID

As its name suggests it is the branch of applied mechanics concerned with the statics and dynamics of fluids - both liquids and gases. The analysis of the behaviour of fluids is based on the fundamental laws of mechanics which relate continuity of mass and energy with force and momentum together with the familiar solid mechanics properties.

We normally recognise three states of matter: solid; liquid and gas. However, liquid and gas are both fluids: in contrast to solids they lack the ability to resist deformation. Because a fluid cannot resist the deformation force, it moves, it flows under the action of the force. Its shape will change continuously as long as the force is applied. A solid can resist a deformation force while at rest, this force may cause some displacement but the solid does not continue to move indefinitely.

The deformation is caused by shearing forces which act tangentially to a surface. Referring to the figure below, we see the force F acting tangentially on a rectangular (solid lined) element ABDC. This is a shearing force and produces the (dashed lined) rhombus element A'B'DC.

$$\sigma = \frac{\rho sub\tan ce}{\rho H_2O(at\ 4°C)}$$

We can then say:

A Fluid is a substance which deforms continuously, or flows, when subjected to shearing forces.

and conversely this definition implies the very important point that:

If a fluid is at rest there are no shearing forces acting. All forces must be perpendicular to the planes which the are acting.

When a fluid is in motion shear stresses are developed if the particles of the fluid move relative to one another. When this happens adjacent particles have different velocities. If fluid velocity is the same at every point then there is no shear stress produced: the particles have zero relative velocity. Consider the flow in a pipe in which water is flowing. At the pipe wall the velocity of the water will be zero. The velocity will increase as we move toward the centre of the pipe. This change in velocity across the direction of flow is known as velocity profile and shown graphically in the figure below:

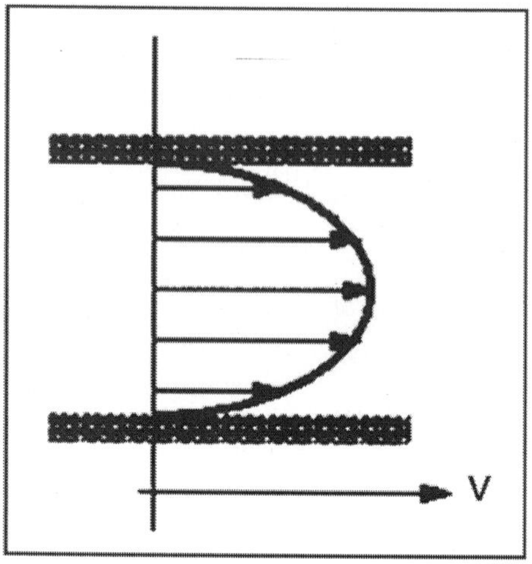

Fig.3.1 Velocity Profile in a Pipe.

Because particles of fluid next to each other are moving with different velocities there are shear forces in the moving fluid i.e. shear forces are normally present in a moving fluid. On the other hand, if a fluid is a long way from the boundary and all the particles are travelling with the same velocity, the velocity profile would look something like this:

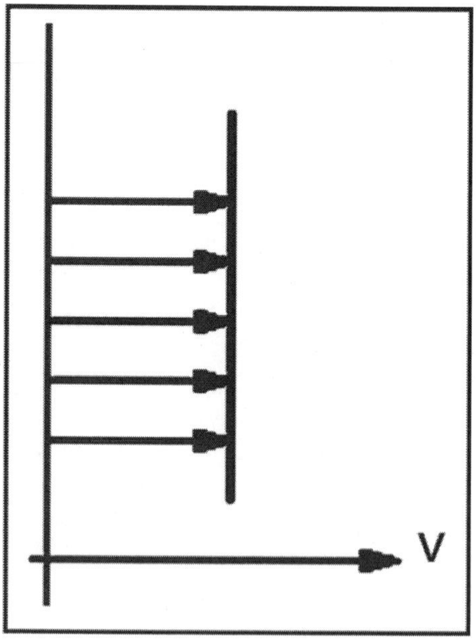

Fig.3.2 Velocity Profile in Uniform Flow

and there will be no shear forces present as all particles have zero relative velocity. In practice we are concerned with flow past solid boundaries; aeroplanes, cars, pipe walls, river channels etc. and shear forces will be present.

Newton's Law of Viscosity

1. Newton's law of viscosity: How can we make use of these observations? We can start by considering a 3d rectangular element of fluid, like that in the figure below.

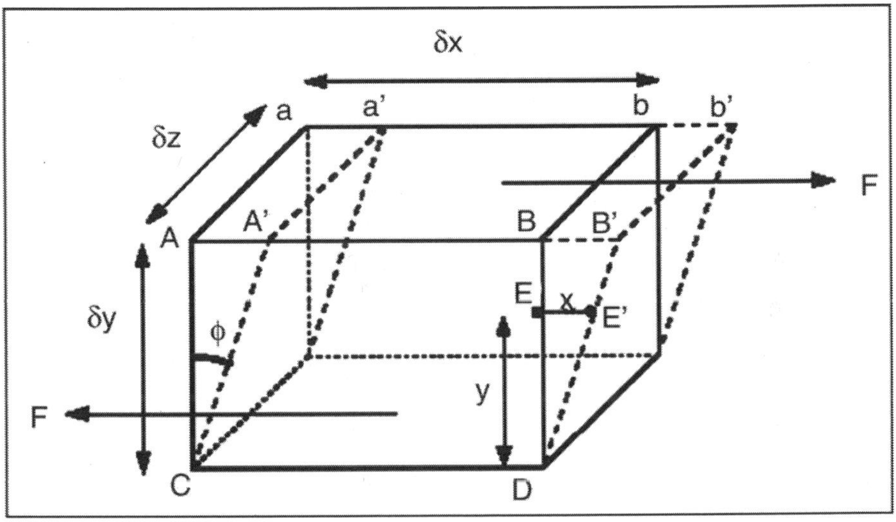

Fig.3.4 Fluid Element Under a Shear Force

The shearing force F acts on the area on the top of the element. We can thus calculate the shear stress which is equal to force per unit area i.e.

shear stress, —

The deformation which this shear stress causes is measured by the size of the angle *f* and is know as *shear strain.*

In a solid shear strain, f, is constant for a fixed shear stress t.

In a fluid f increases for as long as t is applied - the fluid flows.

It has been found experimentally that the rate of shear stress (shear stress per unit time, t/time) is directly proportional to the shear stress. If the particle at point E moves under the shear stress to point E′ and it takes time t to get there, it has moved the distance x.

Using the experimental result that shear stress is proportional to rate of shear strain then,

$$\tau = \text{constant} \times \frac{u}{y}$$

The term,

$$\frac{u}{y}$$

is the change in velocity with y, or the velocity gradient, and may be written in the differential form,

$$\frac{du}{dy}$$

The constant of proportionality is known as the dynamic viscosity, μ, of the fluid, giving

$$\tau = \mu \frac{du}{dy}$$

This is known as Newton's law of viscosity.

2. *Fluids vs. Solids:* Earlier we have discussed the differences between the behaviour of solids and fluids under an applied force. Summarising, we have;
    - For a solid the strain is a function of the applied stress (providing that the elastic limit has not been reached). For a fluid, the rate of strain is proportional to the applied stress.
    - The strain in a solid is independent of the time over which the force is applied and (if the elastic limit is not reached) the deformation disappears when the force is removed. A fluid continues to flow for as long as the force is applied and will not recover its original form when the force is removed.

    It is usually quite simple to classify substances as either solid or liquid. Some substances, however, (e.g. pitch or glass) appear solid under their own weight. Pitch will, although appearing solid at room temperature, deform and spread out over days - rather than the fraction of a second it would take water.

    Looking at properties of solids, when the elastic limit is reached they seem to flow. They become plastic. They still do not meet the definition of true fluids as they will only flow after a certain minimum shear stress is attained.

3. *Newtonian/ Non-Newtonian Fluids:* Even among fluids which are accepted as fluids there can be wide differences in behaviour under stress. Fluids obeying Newton's law where the value of m is constant are known as Newtonian fluids. If m is constant the shear stress is linearly dependent on velocity gradient. This is true for most common fluids.

Fluids in which the value of m is not constant are known as non-Newtonian fluids. There are several categories of these, and they are outlined briefly below. These categories are based on the relationship between shear stress and the velocity gradient (rate of shear strain) in the fluid. These relationships can be seen in the graph below for several categories.

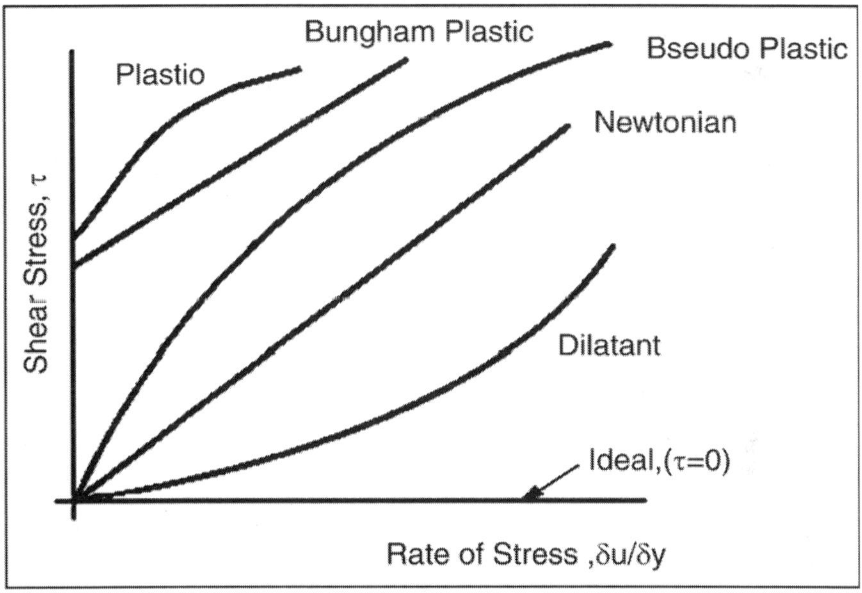

Fig.3.5 Shear Stress vs. Rate of Shear Strain du/dy

Below are brief description of the physical properties of the several categories:

- *Plastic:* Shear stress must reach a certain minimum before flow commences.
- *Bingham plastic:* As with the plastic above a minimum shear stress must be achieved. With this classification n = 1. An example is sewage sludge.
- *Pseudo-plastic:* No minimum shear stress necessary and the viscosity decreases with rate of shear, e.g. colloidial substances like clay, milk and cement.

- *Dilatant substances;* Viscosity increases with rate of shear e.g. quicksand.
- *Thixotropic substances:* Viscosity decreases with length of time shear force is applied e.g. thixotropic jelly paints.
- *Rheopectic substances:* Viscosity increases with length of time shear force is applied
- *Viscoelastic materials:* Similar to Newtonian but if there is a sudden large change in shear they behave like plastic.

There is also one more - which is not real, it does not exist - known as the ideal fluid. This is a fluid which is assumed to have no viscosity. This is a useful concept when theoretical solutions are being considered - it does help achieve some practically useful solutions.

4. Liquids vs. Gasses: Although liquids and gasses behave in much the same way and share many similar characteristics, they also possess distinct characteristics of their own. Specifically,
    - A liquid is difficult to compress and often regarded as being incompressible. A gas is easily to compress and usually treated as such - it changes volume with pressure.
    - A given mass of liquid occupies a given volume and will occupy the container it is in and form a free surface (if the container is of a larger volume).

    A gas has no fixed volume, it changes volume to expand to fill the containing vessel. It will completely fill the vessel so no free surface is formed.
5. *Causes of Viscosity in Fluids:*
    - *Viscosity in Gasses:* The molecules of gasses are only weakly kept in position by molecular cohesion (as they are so far apart). As adjacent layers move by each other there is a continuous exchange of molecules. Molecules of a slower layer move to faster layers causing a drag, while molecules moving the other way exert an acceleration force. Mathematical considerations of this momentum exchange can lead to Newton law of viscosity.

    If temperature of a gas increases the momentum exchange between layers will increase thus increasing viscosity.

    Viscosity will also change with pressure - but under normal conditions this change is negligible in gasses.
    - *Viscosity in Liquids:* There is some molecular interchange between adjacent layers in liquids - but as the molecules are so much closer than in gasses the cohesive forces hold the molecules in place much more rigidly. This cohesion plays

an important roll in the viscosity of liquids.

Increasing the temperature of a fluid reduces the cohesive forces and increases the molecular interchange. Reducing cohesive forces reduces shear stress, while increasing molecular interchange increases shear stress.

High pressure can also change the viscosity of a liquid. As pressure increases the relative movement of molecules requires more energy hence viscosity increases.

SOLIDS AND FLUIDS

All substances of our natural and technical environment can be subdivided into solid, liquid and gaseous media, on the basis of aggregation. This sub- division is considered in many fields of engineering in order to point out important differences concerning the properties of the substances. This could also be applied to fluid mechanics, however, this would not be particularly advantageous. It is rather recommended to employ fluid mechanics aspects to achieve a subdivision of media appropriate for the treatment of fluid flow processes. To this end, the term fluid is introduced for designating all those substances that cannot be classified clearly as solids.

From the point of view of fluid mechanics, all media can be subdivided into solids and fluids, the difference between both groups being that solids possess elasticity as an important property, while fluids have viscosity as a characteristic property. Shear stresses imposed on to a solid from outside lead to inner elastic shear forces which prevent irreversible changes of position of molecules of the solid. When, on the contrary, external shear stresses are imposed on to fluids, they react with the build-up of velocity gradients, the build-up of the gradient occurring through the molecule-dependent momentum transport, i.e. through the fluid viscosity. Thus elasticity (solids) and viscosity (liquids) are the properties of matters that are employed in fluid mechanics for subdividing media. However there are few exceptions to this subdivision: such as in the case of some of the matters in rheology exhibit mixed properties to such an extent that for small deformations they behave like solids and behave like liquids in the case of large deformations.

At this point, attention is drawn to another important fact regarding the characterisation of fluid properties. A fluid tries to evade smallest external shear stresses by starting to flow. Hence it can be inferred that a fluid at rest is characterized by a state which is free of external shear stresses. Each area in a fluid at rest can therefore experience normal forces only. When shear stresses occur in a medium at rest, this medium is assigned to solids. The viscous forces accompanied by external motion observed in a fluid should not be mistaken with the elastic forces in solids. The viscous force cannot be analogously addressed as the elastic force. This is the case for all liquids and gases which take part in fluid motion. The present book is dedicated to such a treatment of fluid flows. On the basis of the above mentioned treatments of fluid flows, the fluids in motion can simply be classified as media free from stresses and distinguished from solids. The "shear stresses" that are often introduced when treating fluid flows of common liquids and gases represent molecule-dependent momentum-transport terms in reality.

Neighboring layers of a flowing fluid, having a velocity gradient, do not interact with each another through "shear stresses" but through an exchange of momentum due to the molecular motion. This can be explained by simplified derivations aiming at the physical understanding of the molecular processes. The derivations are carried out for an ideal gas, since they can be understood particularly well in this case. The results from these derivations can therefore not be transferred in all aspects to fluids with more complex properties.

For further subdivision of the fluids, it is recommended to make use of their response to normal stresses or pressure on fluid elements. When a fluid reacts to pressure changes by changing its volume and consequently density, the fluid is called compressible. When no volume or density changes occur with pressure differences, the fluid is regarded as incompressible. Although strictly speaki ng, incompressible fluids do not exist. However, such a subdivision is reasonable and moreover useful and this will also be shown in following derivations of basic fluid mechanics equations. Indeed, this subdivision -distinguishes liquids from gases.

In general, fluids can be further classified into liquid and gases. Liquids and some plastic materials show very small expansion coeffcients, while gases have much larger expansion coeffcients. A comparison of both subgroups of fluids shows that liquids fulfill the condition of incompressibility with a precision that is adequate for the most of flow problems.

On this assumption, the basic equations of fluid mechanics can be simplified, as the following derivations show; in particular the number of equations needed for the general description of fluid flow processes being reduced from 6 to 4. The simplifications of the basic equations for incompressible media allow a considerable reduction in the complexity of the flow solutions in simple and complex geometries, as in the case of problems without heat transfer the energy equation does not have to be solved.

The simplified basic equations of fluid mechanics derived for incompressible media can occasionally also be applied to the flows of compressible fluids, such as cases where the density variations occurring in the entire flow field are small as compared to the fluid density.

For further characterization of a fluid, it is referred to the well-known fact that solids conserve their form, while a fluid volume has no form of its own, but assumes the form of the container in which it is kept. Liquids differ from gases in terms of the area taken by the fluids constituting only part of the container, while the remaining part is either not filled or contains a gas, there exists a free surface between them. Such a surface does not exist when the container is filled only with a gas. The gas takes up the entire container volume.

Finally, it can be concluded that there is a number of media those can only be categorized in a limited way according to the above classification. They are e.g. media that consist of two phase mixtures. These have properties that cannot be classified so easily. This holds also for a number of other media that can, as per the above classification, be assigned neither to the solids nor to the fluids and they start to flow only above a certain value of the "shear stress". Media of this kind would be excluded in this book, so that the above indicated classifications of media into solids and fluids remain valid. Further restrictions to the fluid properties that are applied in dealing with flow problems in this book are clearly indicated in the respective sections.

In this way it should be possible to avoid mistakes that often arise from the derivations of fluid mechanics equations for simplified fluid properties and/or simplified flow cases.

MOLECULAR PROPERTIES AND QUANTITIES OF CONTINUUM MECHANICS

As all matter consists of molecules or aggregations of molecules, all macroscopic properties of matter can be described by molecular properties. Thus it is possible to evaluate all properties of fluids that are of importance for considerations in fluid mechanics linked to properties of molecules, i.e. to describe the macroscopic properties of fluids by molecular properties. However, such a description of the state of matter requires much efforts due to necessary formalism and moreover would be unclear. A molecular-theoretical presentation of fluid properties would hardly be appropriate to supply practice-oriented fluid mechanics information useful for an engineer in easily comprehensible (and also applicable) form. For this reason, it is more advantageous to introduce quantities of continuum mechanics for describing fluid properties.

The connection between continuum mechanics quantities, introduced in fluid mechanics and the molecular properties should be considered as the most important links between the two different ways of description and presentation of fluid properties.

Some state parameters such as density r, pressure P, temperature T are essential for the description of fluid mechanics processes and can be expressed in terms of molecular quantities for ideal gases. From the following derivations one can infer that the "effects" of molecules or molecular properties on fluid elements or control volumes are taken into consideration by introducing the properties, density r, pressure P, temperature T, viscosity μ etc. in an "integral form" and it is sufficient for fluid mechanics considerations. Therefore, continuum mechanics considerations do not neglect the molecular structure of the fluids, but take them into account in integral form, i.e. averaged over several molecules.

The mass per unit volume is called specific density of a matter. For a fluid element this quantity depends on its position in space, and also on time t, so that generally holds. If *n* is considered to be the mean number of the molecules existing per unit volume and with m the mass available per molecule, the following connection holds:

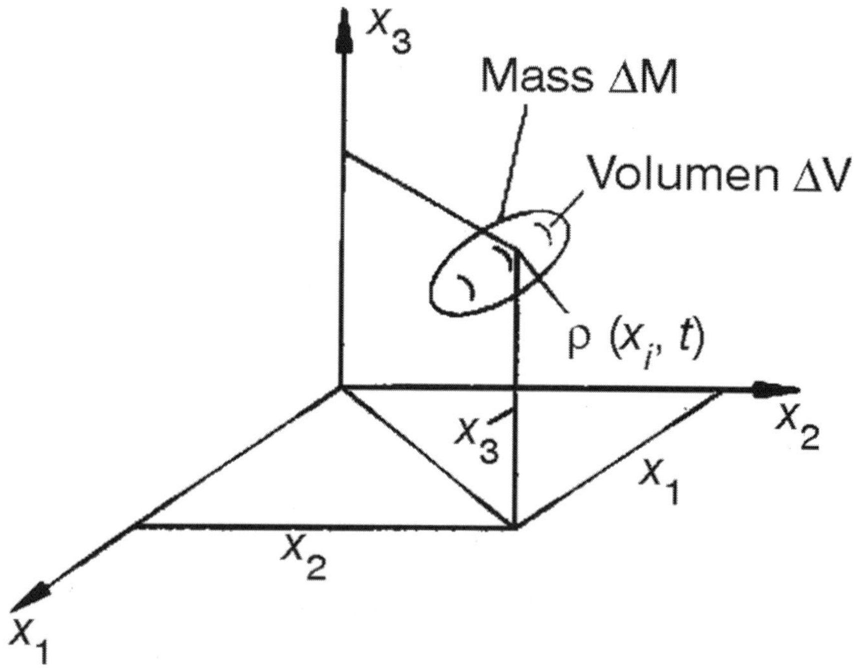

Fig.3.6 Definition of the Fluid Density ρ(xi, t).

The density of the matter is thus identical with the number of molecules available per unit volume, multiplied by the mass of a single molecule. Therefore, changes in density in space and in time correspond to spatial and temporal changes of the mean number of the molecules available per unit volume.

By stochastic consideration of thermal molecular motions in a fluid volume having a large number of molecules under normal conditions, a mean number of molecules can be specified at time t with sufficient clarity. Volumes in the order of magnitude of 10^{-18} – 10^{-20}m3 are considered as sufficiently large for arriving at clear definitions of density. The treatments of flow processes in fluid mechanics are usually carried out in a much larger volume therefore, the specification of "mean" number of molecules in order to designate the available mass in the considered central volume or density is appropriate.

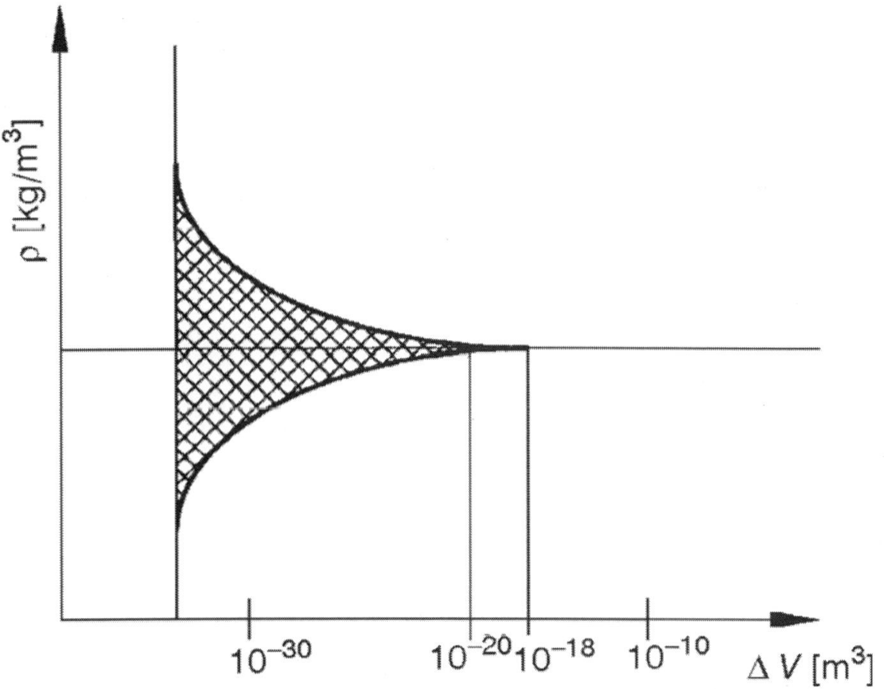

Fig.3.8: Fluctuations while determining the density of fluids.

The local density $\rho(x_i, t)$ therefore describes a property of matter that is essential for fluid mechanics with a precision that is almost always sufficient.

The control volume in the fluid mechanics considerations is selected to such a large extent that the determination of a local density value completely fulfills the requirements of the considerations that are to be carried out from the flow mechanics point of view, in spite of the molecular basic structure of the considered fluids.

Similar considerations can also be made for the pressure that occurs in a fluid at rest and which is defined as the force acting per area unit.

From the molecular-theoretical point of view, the pressure effect is defined as the temporal momentum change occurring per unit area, i.e. the force which the molecules experience and exert when colliding in an elastic way with the considered area.

Analogous to the above volume dimensions, it can be stated that most of the fluid mechanics considerations do not require area resolutions that fall below 10^{-12} bis $10^{-13}m^2$ and therefore the mean numbers of molecules are suffcient to have the force effect of the molecules on the areas.

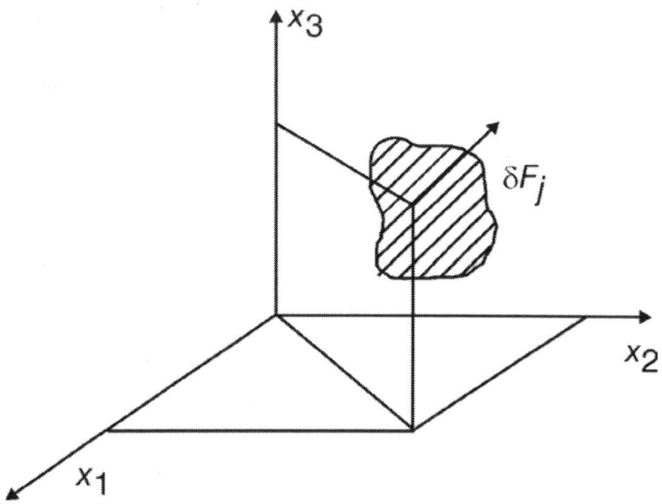

Fig.3.8: Concerning the definition of the pressure in fluid P (xi, t).

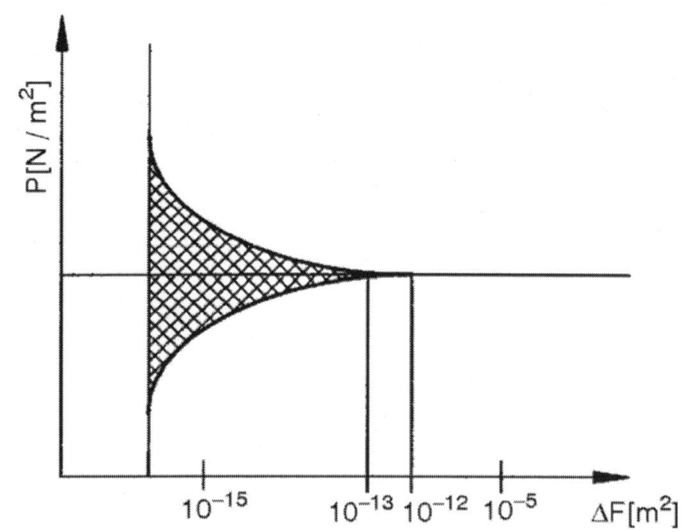

Fig. Fluctuation while determining the pressure in the fluid.

Similar to the above introduced continuum mechanics quantities $\rho(x_i, t)$ and $P(x_i, t)$ there are other local fields such as the temperature, the internal energy and enthalpy of a fluid etc. for which considerations can be repeated analogously to the above indicated treatments regarding the density and the pressure. This again shows that it is sufficient for fluid mechanics considerations to neglect the complex molecular properties and to introduce continuum mechanics quantities into the fluid mechanics considerations that correspond to mean values of corresponding molecular parameters. fluid mechanics considerations can therefore be carried out on the basis of continuum mechanics quantities.

However, there are some important domains in fluid mechanics where continuum considerations are not appropriate, e.g. the investigation of flows in highly diluted gas systems. No clear continuum mechanics quantities can be defined there for the volume and the areas with which fluid mechanics processes are to be resolved, as the required spatial resolution of the flow mechanics considerations does not promise sufficient numbers of molecules for the necessary establishment of the mean values of the parameters which are available with the introduction of the continuum mechanics quantities.

When treating such fluid flows, priority has to be given to the molecular-theoretical considerations of fluid mechanics processes as compared to the continuum mechanics considerations. In the present introduction of the fluid mechanics of viscous media, the domain of flows of highly diluted gases is not dealt with, so that all required considerations can take place in the terminology of continuums mechanics. In these considerations, molecular effects, e.g. within the conservation law for mass, momentum and energy are presented in integral form, i.e. the molecular structure of the considered fluids is not neglected but taken into consideration in the form of integral quantities.

TRANSPORT PROCESSES IN NEWTONIAN FLUIDS

General Considerations

When treating fluids with the transport of heat and molecular mass transport processes occur that cannot be neglected and that hence have to be taken into account in the general transport equations. A physically correct treatment is necessary that orients itself on the general representations and is indicated below. These figures show planes that lie parallel to the $x_1 - x_3$ plane of a cartesian coordinate system. In each of these planes the temperature T = const (a) the concentration c = const (b) and the velocity (U_j) = const (c) are such that when taking into account the increase of the quantities in x_2-direction = x_i- direction, a positive gradient in each of these quantities exists.

The heat transport occurring as a consequence of the molecular motion is given by the Fourier law of heat conduction and the mass transport occurring analogously given by the Fick's law of diffusion.

In an analogous way, the molecule-dependent momentum transport also has to be described by the Newtonian law which in the presence of only one velocity component Uj can be stated as follows here.

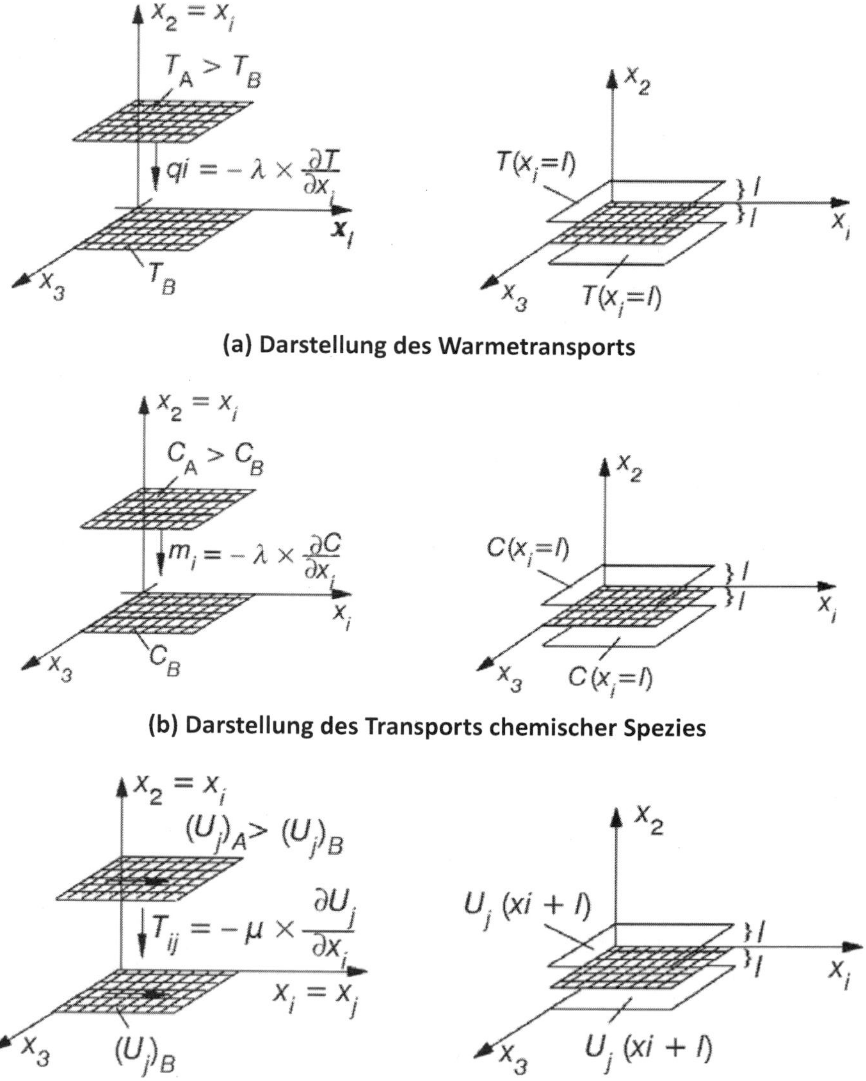

(a) Darstellung des Warmetransports

(b) Darstellung des Transports chemischer Spezies

(c) Darstellung des Impulstransports

Fig.3.10: Analogy of the transport processes dependent on molecules for (a) heat transport, (b) matter transport and (c) momentum transport.

Newtonian law of momentum transport:

μ = dynamic viscosity.

In m_i the direction i indicates the "molecular transport direction", and j indicates the components of the velocity vector for which momentum transport considerations are carried out. The complete equation in the presence of a Newtonian medium can be indicated as follows:

$$\tau_{ij} = \left(\frac{\partial U_j}{\partial x_i} + \frac{\partial U_i}{\partial x_j} \right) + \frac{2}{3} \mu \delta_{ij} \left(\frac{\partial U_k}{\partial x_k} \right)^2$$

τ_{ij} represents momentum transport per unit area for unit time or "stress" i.e. force per unit area. It is therefore often designated as "shear stress" and the sign before μ is chosen positive. This has to be taken into account when comparing representations in this book with corresponding statements in other books. The existing differences in the viewpoints are considered in following two annotations.

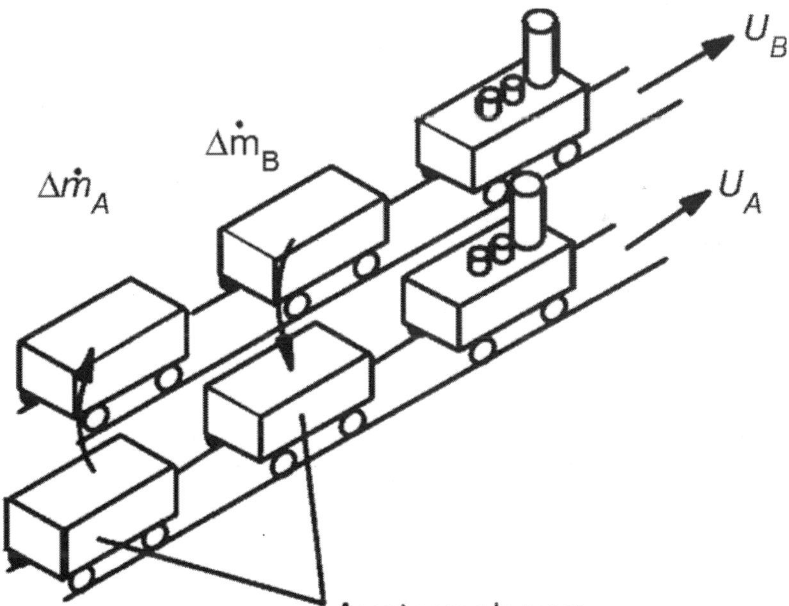

Fig.3.11: Exchange of mass and momentum Illustrative explanation of tij as momentum transport.

Annotation 1: The following illustrative example shall show how the viscosity-dependent momentum transport introduced in the continuum me- chanics is reflected through the motion of molecules. Two passenger trains may run next to one another with different speeds. In each of the trains, persons carry sacks along with them. These sacks are being thrown by the passengers of the one train to the passengers in the other train, so that a momentum transfer takes place; it should be noted that the masses mA and mB of the trains do not change. Due to the fact that the persons in the quicker train catch the sacks that are being thrown to them from the slower train, the quicker train is slowed down. In an analogous way the slower train is accelerated. Momentum transfer in the direction of travel takes place by an momentum transport perpendicular to the direction of travel. This idea, transferred to the molecule-dependent momentum transport in fluids, is in accordance with the molecule dependent transport processes that were stated above.

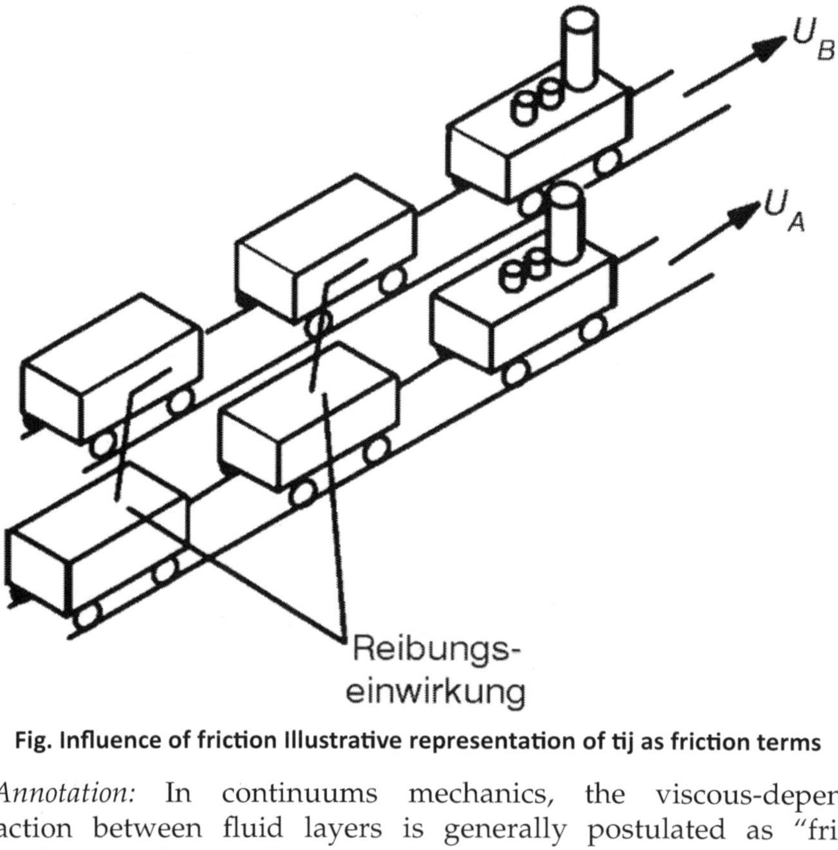

Fig. Influence of friction Illustrative representation of tij as friction terms

Annotation: In continuums mechanics, the viscous-dependent interaction between fluid layers is generally postulated as "friction forces" between layers. This would, in the above described interaction between trains running along each another, correspond to a slowdown or acceleration by frictional forces that could be applied for instance in such a way that the passengers in the trains exert an influence on the motion of the respective other train by bars with which the friction forces along the external wagon wall are induced. This idea does not correspond to the conception of molecular dependent transport processes between fluid layers of different speeds.

If one carries out physically correct considerations regarding the molecular dependent momentum transport tij. In addition, considerations are presented on the following pages concerning pressure, heat exchange and diffusion in gases in order to show the connection between molecular and continuum- mechanics quantities.

PRESSURE IN GASES

From the molecular-theoretical point of view, the gaseous state of aggregation of a matter is characterized by a free or random motion of the atoms and molecules. These laws result from derivations that are based on basic mechanical laws and that start from ideal elastic collisions with which the molecules interact among each another and with walls, e.g. with container walls. Between these collisions, the molecules move freely and in straight lines. This is to say that no forces act between the molecules, except when the collisions take place. Likewise container walls neither attract nor repel the molecules and the interactions of the walls with the moving molecules are limited to the moment of the collision. The most important properties of an ideal gas can be stated as follows:

- The volume of the molecules and the atoms is extremely small as com- pared to their distance from one another so that the molecules can be regarded as material points;
- The molecules exert, except for the moment of the collision, neither at-tractive nor repulsive forces on each another;
- For the collisions between two molecules or a molecule and a wall, the laws of the perfect elastic impact hold. (Collisions of two molecules take place exclusively.)

When one takes into account the characteristic properties of an ideal gas listed in points, the following derivations indicated can be formulated to derive the pressure that represents a characteristic continuum mechanics quantity of the gas, by taking molecular-theoretical considerations into account. These derivations only consider the known basic laws of mechanics and the properties indicated above in points.

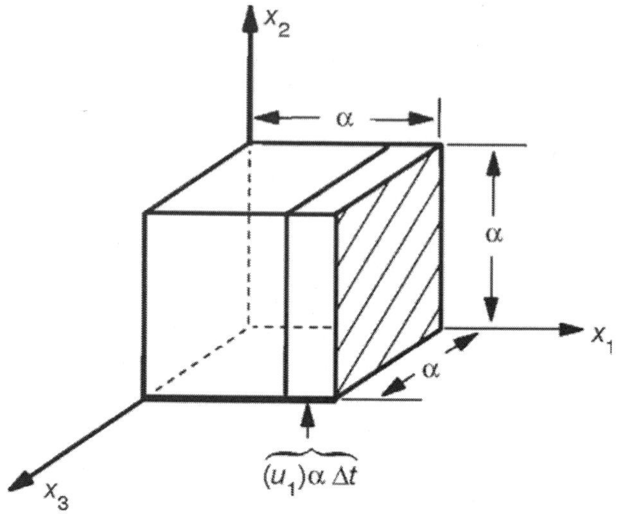

Fig.3.12: Control volume for derivations concerning the pressure effect of molecules

In order to derive the "pressure effect" of the molecules on an area as a consequence of the molecular motion, the derivations carried out by considering a control volume with edge length. Regarding this control volume, the area standing perpendicular to the axis x_1 is shown shaded. All considerations are made for this area. For other areas of the control volume derivations have to be carried out in an analogous way, so that the considerations for the shaded area in figure can be considered as generally valid. In the control volume, N molecules are present altogether. By the introduction of the number of molecules per m^3 (molecular density) n, this number N is given by:

$$N = na^3$$

From $n_α$ molecules per m^3, n_i molecules with a velocity component $(u_1)_i$ may move in the direction of the axis x_1 and interact with the shaded area. In time Δt all molecules will hit the wall area, which have a distance of $(u_1)αΔt$ from it. These are:

$$z_α = n_α a^2(u_1)_α Dt.$$

Each of $z_α$ molecules exerts a momentum on the wall that is formulated by the law of the ideal elastic impact:

$$Δ(i_1)_α = -mD(u_1)_α = 2m(u_1)_α.$$

For the total momentum transferred by za molecules to the wall, it holds:

$$\Delta(J1\,)_\alpha \;= z_\alpha \Delta(i1\,)_\alpha = n_\alpha a_2(u_1)i\Delta t[2m(u_1)_\alpha],$$

$$\Delta(J1)_\alpha = 2m_\alpha 2\Delta t n_\alpha\,(u21)_\alpha.$$

The total pressure which is exerted on the shaded area, summarizes the pressure shares $(P_1)_\alpha$ of all differing velocities $(u_1)_i$. If one wants to calculate the total pressure, one has to summarize over all these contributions.

The summation occurring in the above relation can be substituted by the following definition of the mean value of the velocity square, u^2_1:

MOLECULAR-DEPENDENT MOMENTUM TRANSPORT

Transport processes that are caused by the thermal motion of the molecules were referred in an introductory way. Attention was drawn to the analogy between momentum heat and mass transport and it was pointed out that the τ_{ij} terms used in fluid-mechanics are not considered to be as caused by friction, i.e. physically they represent no "shear stress" but molecular-dependent momentum transports occurring per unit area and time, the index *i* representing the considered transport direction and j is the direction of the considered momentum. In order to give an introduction into a physically correct consideration of the molecular-dependent momentum transport, the below indicated considerations for an ideal gas are made, where only a x_{1-} momentum transport in the direction x_2 is considered, i.e. the term τ_{21}.

For the following derivations a velocity distribution has to be used that does not correspond to the equilibrium distribution (Maxwell distribution), as with this distribution $\tau_{ij} = 0$ and $q_j = 0$.

Therefore the following simple model of a non-equilibrium distribution is used. 1/6 of all molecules at a time moves with a velocity of $(-u\,,0,0)$, $(u,0,0)$, $(0,-u\,,0)$, $(0,u\,,0)$, $(0,0,-u\,)$, $(0,0,u\,)$ with the amount u in parallel to the axis of the coordinates.

When one assumes a molecular concentration per unit volume, i.e. n molecules per m^3, one third of them on an average move with a velocity of u in direction x_2 and of these again one half, i.e. each n/6 molecules per unit volume, move in negative and positive direction x_2. On an average 3 nu molecules per time and unit area unit traverse through the area of plane x_2 = constant, which is indicated in figure.

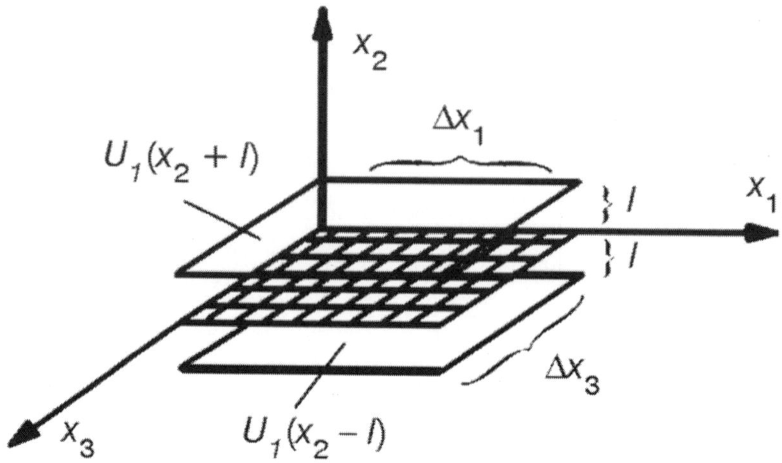

Fig. Molecular motion and sear stress

The molecules which traverse the plane x_2 = const in positive x_2 direction have on an average collided the last time with a distance of l with molecules below the plane, where l represents the mean free path of the molecular motion.

The molecules coming from below thus possess on an average the mean velocity, which the flowing medium has in the plane $(x_2 - l)$. Consequently, the molecular transport in the positive direction x_2 an x_1 momentum, which on averaging can be stated as follows:

$$\Delta J_1 = \frac{1}{6} nu\left[mU_1\left(x_2 - l\right)\right]\Delta t \Delta x_1 \Delta x_3$$

This is connected to an effect of forces τ_{21} quantifiable per time and unit area, i.e. with an effect of forces arising as a consequence of a x_1 momentum that comes about owing to a molecular stream in the positive direction x_2

PHYSICAL PROPERTIES OF FLUIDS

Introduction

A treatment of fluid flow requires an understanding of the physical properties of a fluid which affect the motion, and an appreciation of the units in which these properties are measured.

The three most important fluid properties are the *density, viscosity* and *surface tension*. Density is a familiar concept and is treated briefly, although there is one subtlety that requires attention.

Viscosity is a concept that is generally understood intuitively, but some detailed consideration is necessary in order to establish a precise and useful definition. Each of these will be defined and viewed in terms of molecular concepts, and their dimensions will be examined in terms of mass, length and time (M, L and T).

DENSITY

The *density* p of a fluid is defined as its mass per unit volume and indicates its inertia or resistance to an accelerating force.

Thus,

$$\rho = \frac{mass}{volume} [=] \frac{M}{L^3}$$

in which the notation "[=]" is consistently used to indicate the *dimensions* of a quantity. It is usually understood in Eqn. 2.1 that the volume is chosen so that it is neither so small that it has no chance of containing a representative selection of molecules, nor is it so large that (in the case of gases) changes of pressure cause significant changes of density throughout the volume. A medium characterized by a density is called a *continuum*, and follows the classical laws of mechanics – including Newton's law of motion.

Densities of liquids

Density depends on the mass of an individual molecule and the number of such molecules that occupy a unit of volume. For liquids, density depends primarily on the particular liquid and, to a much smaller extent, on its temperature.

Representative densities are given in table. The accuracy of the values given in tables below through is adequate for the calculation needed for this course.

However, if highly accurate values are needed, particularly at extreme conditions, then specialized information should be sought elsewhere.

Degrees A.P.I. (American Petroleum Institute) are related to specific gravity *s* by the formula:

$$°A.P.I. = \frac{141.5}{s} - 131.5$$

Note that for water, °A.P.I. = 10, with correspondingly higher values for liquids that are less dense. Thus, for the crude oil listed in table, Eqn. indeed gives 141.5/ 0.851 – 131.5 = 35 °A.P.I.

Table. Specific Gravity, Densities, and Thermal Expansion Coefficients of Liquids at 20° C

Liquid	Sp. Gr.	Density, ρ	α Ibm/ft3	8 kg/m3 °C–1
Acetone	0.792	792	49.4	0.00149
Benzene	0.879	879	54.9	0.00124
Crude oil, 35°API	0.851	851	53.1	0.00074
Ethanol	0.789	789	49.3	0.00112
Glycerol	1.26 (50°C)	1,260	78.7	—
Kerosene	0.819	819	51.1	0.00093
Mercury	13.55	13,550	845.9	0.0000182
Methanol	0.792	792	49.4	0.00120
n-Octane	0.703	703	43.9	—
n-Pentane	0.630	630	39.3	0.00161
Water	0.998	998	62.3	0.000207

DENSITY OF GASES

For ideal gases, ρV = *n*RT, where p is the absolute pressure, V is the volume of the gas, *n* is the number of moles, R is the gas constant, and T is the absolute temperature.

If M_w is the molecular weight of the gas, it follows:

$$\rho = \frac{nM_w}{V} = \frac{M_w P}{RT}$$

Thus, the density of an ideal gas depends on the molecular weight, absolute pressure, and absolute temperature. Values of the gas constant R are given in table for various systems of units. Note the degrees Kelvin, formerly represented by "°K" is now more simply donated by "K".

Table. Volume of the Gas Constant, R

Value	Units
8.314	J/g-mole K
0.08314	liter bar/g-mole K
0.08206	liter atm/g-mole K
1.987	cal/g-mole K
10.73	psia ft_3/lb-mole °R
0.7302	ft_3 atm/lb-mole °R
1,545	ft lb_f/lb-mole °R

For a *nonideal gas*, the compressibility factor Z (a function of p and T) is introduced into the denominator of Eqn. giving:

$$\rho = \frac{{}_{n}M_{w}}{V} = \frac{M_{w}P}{RT}$$

Thus, the extent to which Z deviates from unity gives a measure of the nonideality of the gas.

The isothermal compressibility of a gas in defined as:

$$\beta = -\frac{1}{V}\left(\frac{\partial V}{\partial p}\right)_{T}$$

and equals – at constant temperature – the fractional decrease in volume caused by a unit increase in the pressure. For an ideal gas, $\beta = 1/p$, the reciprocal of the absolute pressure.

The *coefficient of thermal expansion* á of a material is its isobaric fractional increase in volume per unit rise in temperature:

$$\alpha = \frac{1}{V}\left(\frac{\partial V}{\partial T}\right)_{P}$$

Since, for a given mass, density is inversely proportional to volume, it follows that for moderate temperature ranges (over which α is essentially constant) the density of most liquids is approximately a linear function of temperature:

$$\rho = \rho_0 \left[1-\alpha\left(T-T_0\right)\right]$$

where ρ_0 is the density at a reference temperature T_0. For an ideal gas, $\alpha = 1/T$, the reciprocal of the absolute temperature.

The *specific gravity* s of a fluid is the ratio of the density ρ to the density ρ_{sc} of a reference fluid at some standard condition:

$$s = \frac{\rho}{\rho_{sc}}$$

For liquids, ρ_{sc} is usually the density of water at 4°C, which equals 1.000 g/ml or 1,000 kg/m^3. For gases, ρ_{sc} is sometimes taken as the density of air at 60°. Fand 14.7 psia, which is approximately 0.0759 lbm/ft$_3$, and sometimes at 0°C and one atmosphere absolute; since there is no single standard for gases, care must obviously be taken when interpreting published values. For natural gas, consisting primarily of methane and other hydrocarbons, the gas gravity is defined as the ratio of the molecular weight of the gas to that of air (28.8 lbm/ lb-mole). Values of the molecular weight Mw are listed in the table below for several commonly occurring gases, together with their densities at standard conditions of atmospheric pressure and 0°C.

Table. Gas Molecular Weights and Densities
The Latter at Atmospheric Pressure and 0°C

Gas	Mw	Standared	Density
Air	28.8	1.29	0.0802
Carbon dioxide	44.0	1.96	0.1225
Ethylene	28.0	1.25	0.0780
Hydrogen	2.0	0.089	0.0056
Methane	16.0	0.714	0.0446
Nitrogen	28.0	1.25	0.0780
Oxygen	32.0	1.43	0.0891

CONTINUUM HYPOTHESIS

The density provides a convenient means of illustrating the continuum hypothesis, which is one of the fundamental principles utilized in fluid mechanics. Density is defined as mass per unit volume, and it is commonly determined by measuring either the mass of a known volume or the volume displaced by a known mass. We recognize, however that density can vary with position in a process, and a more precise definition is to consider a series of small volumes Δv, with corresponding small masses ΔM, and to define the density ρ by the limiting process.

Alternatively, if the density ρ is a function of position, ρ (x, y, z), then the mass contained in a specified volume is the integral of the density over that volume:

$$M = \int \rho\,(x, y, z)\,dn = \iiint \rho\,(x, y, z)\,dx\,dy\,dz$$

Now, Eqns have implicitly used the continuum hypothesis, for they assume that the density is a concept that has meaning for any volume, no matter how small. Yet we know that when the volume is of molecular dimensions we have very rapid spatial variations between dense and non-dense regions, depending on how close we are to a molecule. Hence, the notion of a density function that can be defined by a limiting process such as that of Eqn. as above is questionable. *The continuum hypothesis assumes that mathematical limits for volumes tending to zero are reached over a scale that remains large compared to molecular dimensions.*

Thus, variations on a molecular scale can be ignored, since they average out over all length scales of interest. Clearly, no result derived using the continuum hypothesis should be expected to apply a rarified gas, where the distance between molecules may be large.

VISCOSITY

The viscosity is a measure of resistance to flow. It we tip over a glass of water on the table, the water will spill out before we can stop it.

If we tip over a jar of honey, we probably can set it upright again before much honey flow out; this is because the honey has much greater resistance to flow, more viscosity, than water. A more precise definition of viscosity is possible in terms of following experiment: Consider two long, solid plates separated by a thin film of fluid. If we slide the upper plate steadily in the x direction with velocity Vo, a force is needed to overcome the friction in the fluid between the plates. This force is different for different velocities, plate sizes, fluids, and distances between the plates. We can eliminate the effect of different plate sizes, however, by measuring the force per unit area of the plate, which we define as the shear stress τ.

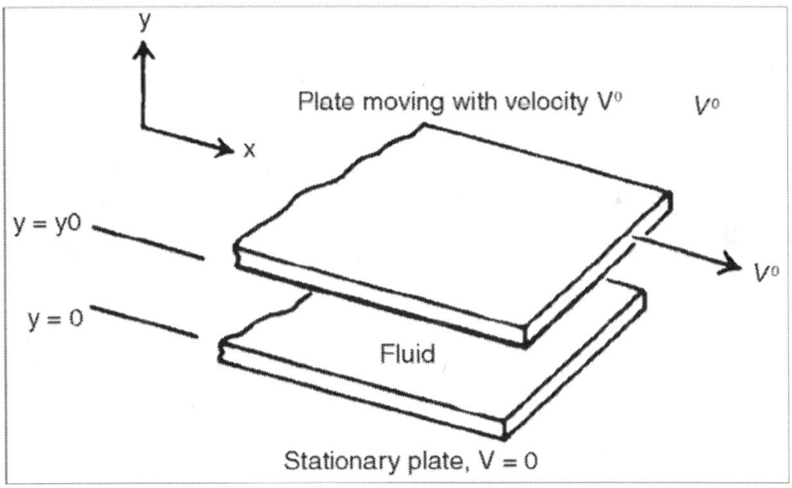

Fig.3.16 The Sliding Plate Experiment.

It has been demonstrated experimentally that at low values of V_0 the velocity profile in the fluid between the plates is linear, i.e.,

$$V = \frac{V_0 y}{y_0}$$

so that,

$$\frac{dV}{dy} = \frac{V_0}{y_0}$$

It has also been demonstrated experimentally that for most fluids the results of this experiment can be shown most conveniently on a plot τ versus dV/ dy. As shown here, dV/ dy is simply is velocity divided by a distance. In more complex geometries, it is the limiting value of such a ratio at a point. It is commonly called the *shear rate, rate of strain,* and *rate of shear deformation,* which all means exactly the same thing. Four different kinds of curve are shown as experiment results in the figure.

All four are observed in nature. The behaviour most common in nature is that represented by the straight line through the origin. This line is called *Newtonian* because it is described by Newton's law of viscosity,

$$\iota = \upsilon \frac{dV}{dy}$$

This equation says that the shear stress τ is linearly proportional to the velocity gradient dV/ dy. It is also the definition of viscosity, because we can rearrange it as,

$$\upsilon \frac{\iota}{dV / dy}$$

Here μ is called the viscosity or the coefficient of viscosity.

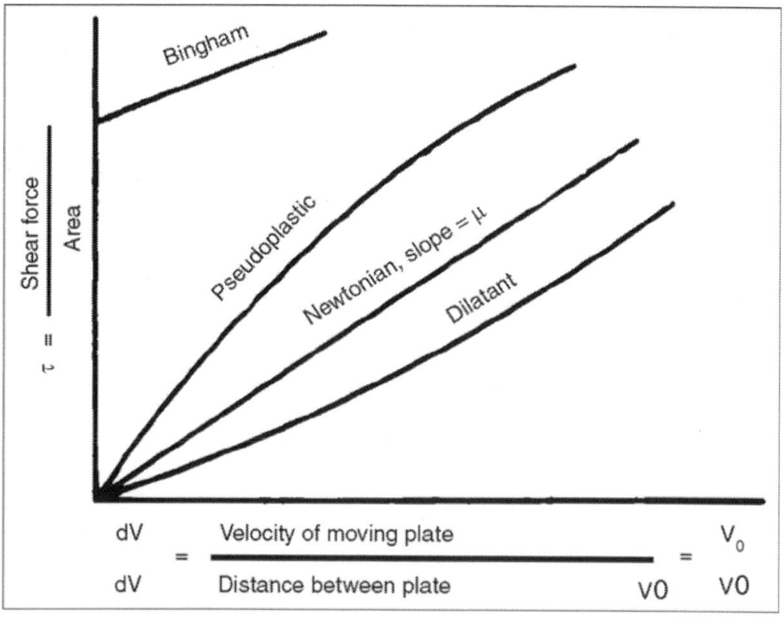

Fig. Possible outcomes of the Sliding-plate Experiment at Constant Temperature and Pressure.

For fluids such as air, the value of μ is very low; therefore, their observed behaviour is represented in Fig above by a straight line through the origin, very close to the dV/dy axis. For fluid such as corn syrup, the value of μ is very large, and the straight line through the origin is close to the τ axis.

Fluids that exhibit this behaviour in the sliding plate experiment (i.e., fluids that obey Newton's law of viscosity) are called *Newtonian fluids.*

All others are called *nonnewtonian fluids*. Which fluids are Newtonian? All gases are Newtonian. All liquids for which we can write a simple chemical formula are Newtonian, such as water, benzene, ethyl alcohol, carbon tetrachloride, and hexane. Most solutions of simple molecules are Newtonian, such as aqueous solutions of inorganic salts and of sugar. Which fluids are nonnewtonian? Generally, nonnewtonian fluids are complex mixtures: slurries, pastes, gels, polymer solutions, etc. Most nonnewtonian fluids are composed of molecules or particles that are much larger than water molecules, such as the sand grains in a mud or the collagen molecules in gelatin, which are thousands or millions of times larger than water molecules.

In discussing non newtonian fluids, we must agree on what we mean by viscosity. If we retain the definition given by Eqn. then the viscosity can no longer be considered a constant independent of *dV/dy* for a given temperature, but must be considered a function of *dV/dy*. This is shown in figure below.

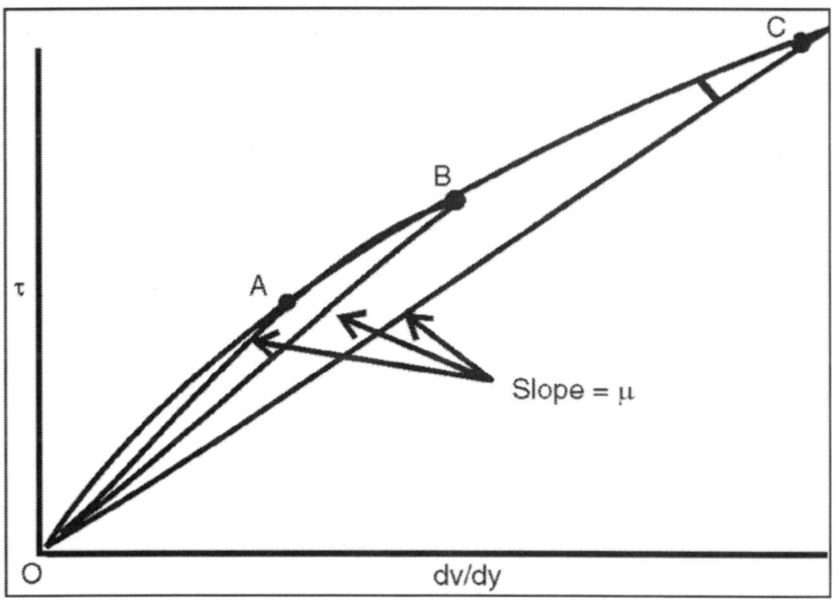

Fig.3.18 Viscosity of Nonnewtonian Fluids at Constant Temperature

Here each of lines OA, OB, and OC have slope μ, so the viscosity is sdecreasing with increasing dV/dy.

Using this definition, we can observe that there are three common types of nonnewtonian fluid:

- *Bingham fluids:* Sometimes called bingham plastics, resist a small shear stress indefinitely but flow easily under larger shear stresses. One may say that at low stresses the viscosity is infinite and that at higher stresses the viscosity decreases with increasing velocity gradient. Examples are bread dough, tooth paste, jellies, and some slurries.
- *Pseudoplastic fluids:* Show a viscosity that decreases with increasing velocity gradient. Examples are most slurries, muds, polymer solutions, solutions of natural gums, and blood.
- *Dilatant fluids:* Show a viscosity that increases with increasing velocity gradient. They are uncommon, but starch suspensions behave in this way.

The basic unit of viscosity is the poise, where 1P = 1g/ (cm.s) = 0.1 Pa.s = 6.72 × 10–2 lbm/ (ft.s). It is widely used for materials such as high-polymer solutions and molten polymers.

However, it is too large a unit for most common fluids. By sheer coincidence the viscosity of pure water at about 68°F is 0.01 P; for that reason the common unit of viscosity in the United States is the centipoises, where

1 cP = 0.01 P = 0.01 g/ (cm.s) = 6.72 × 10–4lbm/ (ft.s) = 0.001Pa.s.

Hence, the viscosity of a fluid, expressed in centipoises, is the same as the ratio of its viscosity to that of water at room temperature.

Kinematic Viscosity

In many engineering problems, viscosity appears only in the relation of viscosity divided by density.

(The kinematic viscosity of water at room temperature is 1cSt.) To avoid confusion over which viscosity is being used, some writers refer to the viscosity μ as the absolute viscosity.

SURFACE TENSION

Surface tension is the tendency of the surface of a liquid to behave like a stretched elastic membrane. There is a natural tendency for liquids to minimize their surface area. The obvious case is that of a liquid droplet on a horizontal surface that is not wetted by the liquid – mercury on glass, or water on a surface that also has a thin oil film on it. For small droplets, such as those on the Fig. the droplet adopts a shape that is almost perfectly spherical, because in this configuration there is the least surface area for a given volume.

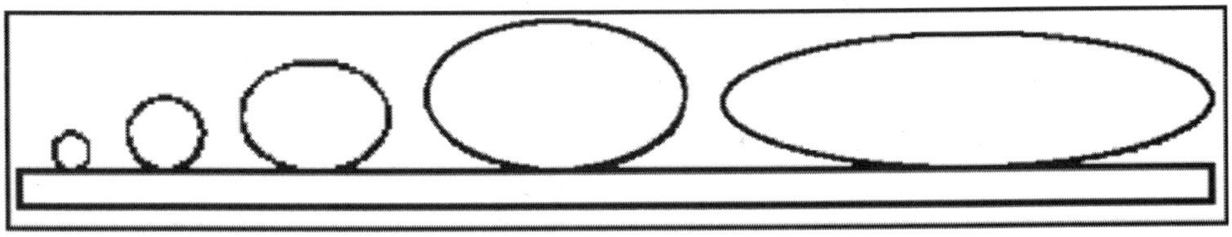

Fig.3.19: The Larger Droplets are Flatter Because Gravity is Becoming more Important than Surface Tension.

For larger droplets, the shape becomes somewhat flatter because of the increasingly important gravitational effect, which is roughly proportional to a3, where a is the approximate droplet radius, whereas the surface area is proportional only to a^2.

Thus, the ratio of gravitational to surface tension effects depends roughly on the value of $a^3/a^2 = a$, and is therefore increasingly important for the larger droplets, as shown to the right in the Figure above.

Overall, the situation is very similar to that of a water-filled balloon, in which the water accounts for the gravitational effect and the balloon acts like the surface tension.

A fundamental property is the surface energy, which is defined with reference to the Figure a. A molecule I, situated in the interior of the liquid, is attracted equally in all directions by its neighbors.

However, a molecule S, situated in the surface, experiences a net attractive force into the bulk of the liquid. (The molecule S.) Therefore, work has to be done against such a force in bringing an interior molecule to the surface. Hence, an energy σ, called the *surface energy,* can be attributed to a unit area of the surface.

Fig. (a) Molecules in the Interior and Surface of a Liquid; (b) Newly Created Surface Caused by Moving the Tension T Through a Distance L.

An equivalent viewpoint is to consider the surface tension T existing per unit distance of a line drawn in the surface, as shown in the Figure b. Suppose that such a tension has moved a distance L, thereby creating an area WL of fresh surface. The work done is the product of the force, TW, and the distance L through which it moves, namely TWL, and this must equal the newly acquired surface energy σWL Therefore, T = σ; both quantities have units of force per unit distance, such as N/m, which is equivalent to energy per unit area, such as J/m^2.

UNITS AND SYSTEMS OF UNITS

Mass, weight, and force

The mass M of an object is a measure of the amount of matter it contains and will be constant, since it depends on the number of constituent molecules and their masses. On the other hand, the *weight w* of the object is the gravitational force on it, and is equal to *Mg*, where *g* is the local gravitational acceleration. Mostly, we shall be discussing phenomena occurring at the surface of the earth, where g is approximately 32.174 ft/s^2 = 9.807 m/s^2= 980.7 cm/s^2. These values are simply taken as 32.2, 9.81 and 981, respectively.

Table. (Representative Units of Force)

System	Units of Force	Customary Name
SI	kg m/s^2	Newton
CGS	g cm/s^2	Dyne
FPS	lbm ft/s^2	Poundal

from which is apparent that force has dimensions ML/T^2. Table gives, the corresponding units of force in the SI (meter/kilogram/second), CGS (centimeter/ gram/ second), and FPS (foot/ pound/ second) systems. The poundal is now an archaic unit, hardly ever used. Instead, the pound force, ibf, is much more common in the English system; it is defined as the gravitational force on 1 ibm, which, if left to fall freely, will do so with an acceleration of 32.2 ft/s^2.

When using lbf in the ft, lbm, s (FPS) system, the following conversion factor, commonly called "gc".

Some writers incorporate gc into their equations, but this approach may be confusing since it virtually implies that one particular set of units is being use, and hence tends to rob the equations of their generality. Why not, for example, also incorporate the conversion factor of 144 in2/ft2 into equations where pressure is expressed in lbf/in2? We prefer to omit all conversion factors in equations, and introduce them only as needed in evaluating expressions numerically. If the reader is in any doubt, units should always be checked when performing calculations.

SI Units: The most systematically developed and universally accepted set of units occurs in the SI units or *Systeme International d'Units;* the subset we mainly need is shown in table.

The basic units are again the meter, kilogram, and second (m, kg, and s); from these, certain *derived* units can also be obtained. Force (kg m/s^2) has already been discussed; energy is the product of force and length; power amounts to energy per unit time; surface tension is energy per unit area or force per unit length, and so on.

Some of the units have names, and these, together with their abbreviations, are also given in table.

Table. (SI Units)

Psysical Quantity	Name of Unit	Symbol for Unit	Definition of Unit
Basic Units			
Length	meter	m	–
Mass	kilogram	kg	–
Time	second	s	–
Temperature	degree		
	Kelvin	K	–
Supplementary unit			
Plane angle	radian	rad	–
Derived Units			
Acceleration			m/s^2 Angular
Velocity			rad/s
Density			kg/m^3
Energy	joule	J	kg/m^3
Force	newton	N	kg/m^2/s^2
Kinematic viscosity			
Power	watt	W	m^2/s^2
Pressure	pascal	Pa	Kg/m^2/s^3 (J/s)
Velocity			m/s
Viscosity			kg/m s

Tradition dies hard, and certain other "metric" units are so well established that they may be used as *auxiliary units;* these are shown in table.

Table. (Auxiliary Units Allowed in Conjunction with SI Units)

Physical Quantity	Name of Unit	Symbol for Unit	Definition of Unit
Area	hectare	ha	10^4 m^2
inematic viscosity	stokes	St	10^{-4} m^2/s
Length	micron	µm	10^{-6} m
Mass	tonne	t	10^3 kg = Mg
	gram	g	10^{-3} kg = g
Pressure	bar	bar	10^5 N/m^2
Viscosity	poise	P	10^{-1} kg/ ms
Volume	liter	1	10^{-3} m^3

Table shows some of the acceptable prefixes that can be used for accommodating both small and large quantities. For example, to avoid an excessive number of decimal places, 0.000001 s is normally better expressed as 1 µs (one microsecond). Note also, for example, that 1 µkg should be written as 1 mg-one prefix being better than two.

Table. (Prefixes for Fractions and Multiples)

Factor	Name	Symbol	Factor	Name	Symbol
10–12	pico	p	10^3	kilo	k
10–9	nano	n	10^6	mega	M
10–6	micro	µ	10^9	giga	G
10–3	milli	m	10^{12}	tera	T

CLASSIFICATION OF FLUID FLOWS

Kinematic Classification Of Fluid Motion

One way of classifying different types of motion is according to their kinematic properties. For example, there is a large class of motions dealing with divergenceless fluids (commonly referred to as motion of incompressible fluids. Another class is motion wherein the vorticity is everywhere zero. This is referred to as irrotational motion. Motions of solid bodies, representing perhaps the most widely studied class in applied mechanics, falls under the heading of *deformationless motion.*

STREAM FUNCTION FOR TWO-DIMENSIONAL FLOW

If a particular two-dimensional motion is divergenceless, implying that,

$$\frac{\partial u}{\partial x}+\frac{\partial v}{\partial y}=0$$

then the velocity field can be represented in terms of a single scalar function commonly referred to as the streamfunction. If we let represent the streamfunction, then

$$u=-\frac{\partial \psi}{\partial y}$$

and

$$v=\frac{\partial \psi}{\partial y}$$

Equation is a necessary condition that such a function as implied by Eq. exist. It is readily verified that equations satisfy the condition, provided that

$$\frac{\partial}{\partial x}\frac{\partial \psi}{\partial y}=ppxy\left(\frac{\partial \psi}{\partial x}\right)$$

which holds as long as µ possesses first and second derivatives.

VELOCITY POTENTIAL

The quantity µ, referred to as the velocity potential, is a function of x,y,z,t in the most general case. Unlike the streamfunction, which is applicable for divergenceless, two-dimensional flow, the velocity potential may exist even for three-dimensional flow. However, the necessary condition for the existence of µ as defined by Eq. is somewhat more restrictive since, in effect, Eq. requires that

$$\frac{\partial w}{\partial y}=\frac{\partial v}{\partial z},\frac{\partial u}{\partial z}=\frac{\partial w}{\partial x},\frac{\partial v}{\partial x}=\frac{\partial u}{\partial y}$$

LAPLACIAN FLOW

A two-dimensional field of flow which is both divergenceless and irrotational can be characterized by either one of the scalar functions μ or μ.

This type of flow, which bears the name of Laplace, possesses a number of interesting features which have been extensively investigated, and forms the bulk of much of the classical literature on the subject of hydrodynamics. Only some of the properties of this type of flow will be dealt with here. For a more exhaustive treatment, the reader is referred to the incomparable work of Lamb.

CONFORMAL MAPPING

It is possible at once to write down the most general solution for □ and □ as governed by the equations of Art.. This can be done by introducing the concept of functions of a complex variable. If we let,

$$W = \phi + \iota\psi$$

and,

$$Z = x + iy$$

then,

$$W = f(Z)$$

represents the most general solution for ϕ and ψ, where f is any differentiable function at all. In any particular case, the problem is reduced to that of finding a function f which will satisfy certain boundary conditions.

The proof that Eq. satisfies the basic conditions on ψ and ϕ expressed by Eqs. is straightforward.

$$u = -\frac{\partial \psi}{\partial y}$$

and

$$v = \frac{\partial \psi}{\partial y}$$

where it is considered that the quantities ϕ and ψ are real. These equations are in accord with euations, as can be seen by equating the two expressions for u and the two expressions for v. The conditions in fact represent the Cauchy- Riemann conditions which are required if the function f(Z) is to possess a unique derivative.

The relation between W and Z stipulated by implies that for each point x, y in the Z-plane there corresponds at least one value of ϕ and ψ in the W-plane. One can map out the isolines of ϕ and ψ in the x, y plane or isolines of x and y in the ϕ and ψ plane. In either case the sets of isolines are orthogonal.

Consider an arbitrary curve C on the Z-plane. This can be mapped on the W-plane through a given transform w = F(z). Suppose the curve C and its mapped counterpart. Any point on the curve in the Z-plane can be represented by x, y or by the polar coordinates r, ϕ and ψ .

The solution of a problem of Laplacian flow in the vicinity of a fixed boundary is tantamount to finding that transform w = F(z) which maps the boundary as the straight line = constant in the W-plane. That is, for a fixed boundary the flow must be along the boundary and consequently the latter must be a surface of constant.

Consider the problem of ascertaining the two-dimensional, irrotational, divergenceless flow in the neighborhood of the solid boundary formed by two planes intersecting at an arbitrary angle.

The boundary, c – c′, can be mapped on the W-plane as a straight line by use of the following transform:

$$W = zN.$$

If we introduce the complex polar form of W and Z given by Eqs., we have from Eq.

$$Re^{i\theta} = r^{n}e^{in}\vartheta$$

which implies that, and,

$$R=z^{N}$$

and

$$\theta = n\vartheta$$

If we then choose the value of n as,

$$n = \frac{\pi}{alpha}$$

then the curve c - c′ will represent the axis of δ on the W plane, i.e. the line $\theta = 0$, which satisfies the condition that the boundary is a streamline.

The streamlines for other positions in the fluid can now be mapped from the W-plane back onto the Z-plane by use of the relations. As a specific example suppose a radians, then,

$$n = 3/2$$

and,

$$r = R^{2/3}$$

$$\vartheta = 2/3\theta$$

For any particular isoline ψ = constant, the value of R consistent with a selected ϑ can be found from the isoline of ϑ in the W-plane and mapped on the Z-plane using the transformed coordinates r, ϑ from equations. The isolines of ϑ can be mapped on the Z-plane in a similar manner. The resulting orthogonal fields of ϑ and ϑ. Either set of isolines determines the flow uniquely in the neighborhood of the boundary. The isolines of ϑ obtained in this illustration represent a family of hyperbolas whose major axis is the isoline $\vartheta = 0$ and whose asymptotes are the boundary lines c – c′.

Hence at point 0 the velocity is zero, i.e. the point of intersection of the straight portions of the boundary represents a stagnation point in the flow. Eqs. indicate that the velocity increases with distance from the boundary, the magnitude simply being,

$$V\sqrt{V_r^2 + V_\vartheta^2} = \frac{3}{2}r^{1/2}$$

Consequently, the solution can only apply over a finite distance from the boundary if the motion is to be physically real (i.e. have a finite velocity at infinity).

A thorough discussion of conformal transformations is beyond the scope of this text. For a more thorough treatment including a wide variety of solved problems of this type, reference is made to Milne-Thompson (1938).

IDEAL FLUIDS

The flow in which effects of tangential or shearing forces are taken into account; these forces give rise to fluid friction, because they oppose the sliding of one particle past another. We have so far considered ideal fluids, which are not only incompressible but also nonviscous.

In the absence of viscosity, mechanical energy is conserved, which is the essence of Bernoulli's equation.

However, to understand one of the more important properties of a fluid, its ability to exert drag forces on objects, we must go beyond the nonviscous approximation and incorporate the effects of viscosity into our description of fluid flow.

VISCOSITY

When we deform a solid, so that it is strained, we know that the solid will exert a restoring force which opposes the strain; for small strains the restoring force is proportional to the strain, and we have the familiar Hooke's Law. Real fluids also oppose strains; however, in a fluid it is not the amount of strain which is important but the rate at which the strain is produced.

For instance, if we have to draw a utensil through our favourite viscous fluid (maple syrup, say), we find that it is easy for slow motions of the utensil, but more difficult if the utensil is moved rapidly. A simple fluid flow is shown in Figure.

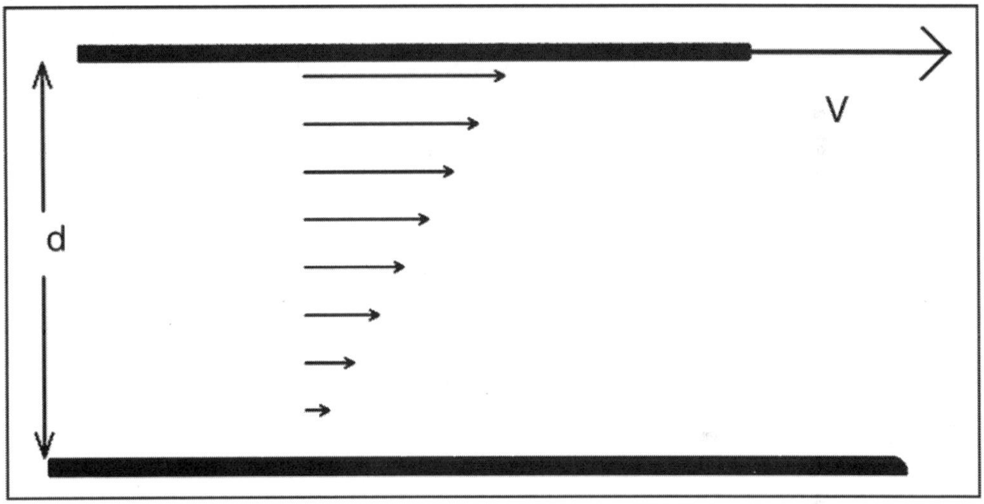

Fig.3.21: Viscous Drag between Two Plates; The Bottom Plate is Held Fixed.

If the force per unit area of the top plate is measured, we would find,

$$\frac{F}{A} = \eta \frac{v}{d}$$

with d the separation between the plates. This essentially defines the viscosity ϑ (sometimes called the shear viscosity). Note that we haven't derived this law—it is a consequence of experimental observation, just like Ohm's law, Fourier's law and Fick's law. A fluid which responds to a shear stress F/A in this manner is called a Newtonian fluid ; it has the property that the viscosity is independent of velocity. Most of the fluids that we will be interested in will be Newtonian (air, water). Non-Newtonian fluids (Silly-Putty, paint, polymeric fluids) are more complicated and beyond the scope of the present description.. Also listed in the kinematic viscosities.

Recall that 1 pascal equals 1 kgm-1s-2 and note that the viscosity of water is, conveniently, 1 mPas = 1 gm-1s-1; it is common to quote relative viscosities, which are viscosities relative to water, or in mPas, and also to use the cgs unit of viscosity, gcm-1s-1, called the poise (after Poiseuille) and equal to 0.1 Pas.

Before incorporating viscosity into the equations of fluid mechanics, let's take a moment to discuss some of the properties of viscosity. Viscosity does not depend upon pressure in a significant way, but we know from personal experience that it does depend upon temperature—pancake syrup becomes less viscous after we take it out of the refrigerator and let it warm up.

Generally, the viscosity of liquids decreases with increasing temperature. Gases, on the other hand, have viscosities which generally increase with increasing temperature. Why the difference?

Viscosity is fundamentally a consequence of the intermolecular interactions in the fluid. Now viscosity is the transfer of momentum from one part of a fluid to an adjacent part, and in a dilute gas this transfer will be more effective if the collisions are very energetic. Since the mean kinetic energy of a molecule is (3/2)kBT, higher temperatures imply a larger viscosity for a gas. A liquid, on the other hand, is a strongly interacting collection of molecules, and the idea of a binary collision is meaningless. Microscopically the molecules in a liquid do have some short range order—a given molecule is surrounded by a "cage" of other molecules.

Momentum transfer in the liquid will involve the movement of these cages around one another. Increasing the temperature causes the cages to jiggle a bit more, allowing them to slip more easily past one another, thereby reducing the momentum transfer between adjacent bits of the fluid, and thereby reducing the viscosity of the liquid.

THE STRESS TENSOR AND THE NAVIER-STOKES EQUATION

The velocity of a fluid will vary in a complicated way in space; however, we can still apply the above definition of viscosity to a bit of fluid of thickness Δy with an infinitesimal area Δy= ΔzΔx. The shear stress across the thickness Δy is,

$$\frac{\Delta F_x}{\Delta A_y} = \eta \frac{\delta v_x}{\delta_y}$$

We can further generalize and define the xy component of the viscous stress tensor as,

$$\tau_{xy} = \eta \left(\frac{\Delta v_x}{\Delta y} + \frac{\partial v_y}{\partial x} \right)$$

The notation is made clearer in the figure below. This definition ensures that for a uniform rotation of the fluid there are no viscous stresses.

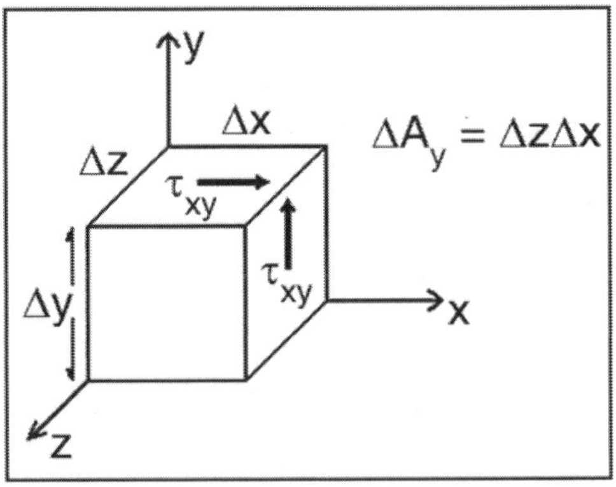

Fig.3.22: Stresses Acting on a Small Cube

Once we have the stress tensor, the force per unit volume of the fluid is obtained by considering the stresses acting upon an infinitesimal cube of the fluid, and calculating the net force on the cube by taking the differences in the stress acting on different faces (just like our calculation of the force arising from pressure gradients in the fluid).

Along with the equation of continuity, it is the basic equation which governs the flow of Newtonian fluids. To complete the description of the flow we need to specify the boundary conditions on the velocity field at a solid surface. In our dealings with nonviscous flow we imposed the condition that the normal component of the fluid velocity at a solid surface must be zero — this just ensures that no fluid flows through the surface. Once we deal with real (viscous) fluids, we must also require that the *tangential* component of the velocity of the fluid be zero on the surface. This condition is often called the no-slip boundary condition, and is the result of experimental observations.

❐

3.

Working in Statistical Physics

Statistical physics is the branch of physics that uses methods of probability theory and statistics, and particularly the mathematical tools for dealing with large populations and approximations, in solving physical problems. It can describe a wide variety of fields with an inherently stochastic nature. Its applications include many problems in the fields of physics, biology, chemistry, neurology, and even some social sciences, such as sociology. Its main purpose is to clarify the properties of matter in aggregate, in terms of physical laws governing atomic motion.

In particular, statistical mechanics develops the phenomenological results of thermodynamics from a probabilistic examination of the underlying microscopic systems. Historically, one of the first topics in physics where statistical methods were applied was the field of mechanics, which is concerned with the motion of particles or objects when subjected to a force.

Statistical physics is an unfinished and highly active part of physics. We do not have the theoretical framework in which to even attempt to describe highly irreversible processes such as fracture. Many types of non-linear systems that lead to complicated pattern formation processes, the properties of granular media, earthquakes, friction and many other systems are beyond our present understanding and theoretical tools.

In the study of these and other macroscopic systems we build on the established conceptual framework of thermodynamics and statistical physics. Thermodynamics, put into its formal and phenomenological form by Clausius, is a theory of impressive range of validity. Thermodynamics describes all systems form classical gases and liquids, through quantum systems such as superconductors and nuclear matter, to black holes and elementary particles in the early universe in exactly the same form as they were originally formulated. Statistical physics, on the other hand gives a rational understanding of Thermodynamics in terms of microscopic particles and their interactions. Statistical physics allows not only the calculation of the temperature dependence of thermodynamics quantities, such as the specific heats of solids for instance, but also of transport properties, the conduction of heat and electricity for example. Moreover, statistical physics in its modern form has given us a complete understanding of second-order phase transitions, and with Wilson's Re-normalisation Group theory we may calculate the scaling exponents observed in experiments on phase transitions.

However, the field of statistical physics is in a phase of rapid change. New ideas and concepts permit a fresh approach to old problems. With new concepts we look for features ignored in previous experiments and find exciting results. Key words are deterministic chaos, fractals, self-organised criticality (SOC), turbulence and intermittency. These words represent large fields of study which have changed how we view Nature.

Disordered systems, percolation theory and fractals find applications not only in physics and engineering but also in economics and other social sciences. What are the processes that describe the evolution of complex systems? Again we have an active field with stochastic processes as a key word. So what has been going on? Why have the mysteries and excitement of statistical physics not been introduced to you before?

The point is that statistical physics is unfinished, subtle, intellectually and mathematically demanding. The activity in statistical physics has philosophical implications that have not been discussed by professional philosophers—and have had little influence on the thinking of the general public.

DETERMINISTIC behaviour rules the day. Privately, on the public arena, in industry, management and in teaching we almost invariably assume that any given action leads predictably to the desired result—and we are surprised and discouraged when it does not.

The next level is that some feed-back is considered—but again things get too simplistic. Engineers who study road traffic reach the conclusion that more roads will make congestions worse, not better—a statement that cannot be generally true. Similarly, the "Club of Rome" report predicted that we would run out of resources very quickly.

The human population would grow to an extent that would outstrip food, metals, oil and other resources. However, this did not happen. Now, with a much large population we have more food, metals, and oil per capita than we had when the report was written. Of course, statistical physics cannot be applied to all areas. The point is that in physics, statistical physics provides us with many examples of complex systems and complex dynamics. Statistical physics provides an intellectual framework and a systematic approach to the study of real world systems with complicated interactions and feedback mechanisms.

SYSTEMS OF ATOMISTIC PHILOSOPHY

Now, for systems with many active degrees of freedom, we observe cooperative phenomena, i.e. behaviour that does not depend on the details of the system. We observe universality, or universal behaviour, that cannot be explained by an atomistic understanding alone. Let me explain: The atomistic philosophy is strong and alive.

We believe that with an atomistic understanding of the particles that constitute matter—and their interactions—it is merely an exercise in applied mathematics to predict the macroscopic behaviour of systems of practical importance. This certainly is an arrogant physicists attitude. It must be said, however, that the experiments and theories are entirely consistent with this view—in spite of deterministic chaos, self-organised criticality and their ilk.

It is clear that the quantum mechanics required to describe the interactions of atoms in chemical systems is fully understood. We have the exact solution for the Hydrogen atom, the approximate solution for the Helium atom, and numerical solutions for more complex atoms and their interactions in molecules. There is no evidence that there are open questions in the foundation of quantum chemistry. Thus we know what is required for a rational understanding of chemistry—the work has been done, we only need to apply what is known practical tasks in engineering and industry so to say.

Since we understand chemistry, no new fundamental laws of Nature are required to understand biochemistry, and finally by implication the brain—how we think and understand! Clearly something is wrong with this line of reasoning. Molecular biology is at present the field of science that has the highest rate of discovery, while physicist have been waiting for almost thirty years for the observation of the Higgs Boson predicted by the standard model Elementary particle theory clearly lies at the foundation of physics—but the rate of discovery has slowed to a glacial pace.

Quantum mechanics was a fantastic break-through, which completely dominated physics in the first half of previous century. Discoveries, were made at an astounding rate. Suddenly we understood spectra, the black-body radiation, X-rays, radioactivity, all kinds of new particles, electrons, protons, neutrons, antiparticles and much more. Much was finished before the second World war, whereas nuclear reactors, and weapons, were largely developed in 50's and 60's.

The search for elementary particles, the constituents of nuclear matter, and their interactions gained momentum in the 50's and truly fantastic research organisations of unprecedented size, such as CERN, dedicated to basic research were set up.

This enthusiasm was appropriate and lead to a long string of discoveries (and Nobel prizes) that have changed our society in fundamental ways. Quantum mechanics is, of course, also at the basis of the electronics industry, computers, communication, lasers and other engineering products that have changed our lives. However, this huge effort on the atomistic side, left small resources—both human and otherwise—to other parts of physics. The intellectual resources were largely concentrated on particle physics research, the other aspects of the atomistic philosophy, namely the understanding of the natural world in terms of the (newly gained) atomistic physics, was left to an uncoordinated little publicized and conceptually demanding effort.

Cooperative Phenomena and Hydrodynamics

Cooperative phenomena are ubiquitous and well known from daily life. Take hydrodynamics for example. Air, water, wine, molten steel, liquid argon — all share the same hydrodynamics. They flow, fill vessels, they from drops and bubbles, have surface waves, vortices and exhibit many other hydrodynamic phenomena—in spite of the fact that molten steel, water and argon are definitely described by quite different Hamilton operators that encode their quantum mechanics.

How can it be that important macroscopic properties do not depend on the details of the Hamiltonian? Phase transitions are also universal: Magnetism and the liquid–vapour transition belong to the same universality class; in an abstract sense they are the same phase transition! Ferro-electricity, spinodal decomposition, superconductivity, are all phenomena that do not depend on the details of their atomistic components and their interactions—they are cooperative phenomena. The understanding of cooperative phenomena is far from complete. We have no general theory, except for second order phase transitions, and the phenomenological equations for hydrodynamics, to classifyand simplify the general scaling and other fractal phenomena observed in highly non-equilibrium systems. Here we have a very active branch of statistical physics.

BOLTZMANN'S WORK IN STATISTICAL PHYSICS

Popular Perceptions of Boltzmann

Boltzmann's work met with mixed reactions during his lifetime, and continue to do so even today. It may be worthwhile, therefore, to devote a few remarks to the perception and reception of his work.

Boltzmann is often portrayed as a staunch defender of the atomic view of matter, at a time when the dominant opinion in the German-speaking physics community, led by influential authors like Mach and Ostwald, disapproved of this view. Indeed, the story goes, in the late nineteenth century any attempt at all to search for a hypothetical, microphysical underpinning of macroscopic phenomena was regarded as suspicious. Further, serious criticism on his work was raised by Loschmidt and Zermelo. Various passages in Boltzmann's writing, especially in the late 1890s, complain that his work was hardly noticed (entitling one article "On some of my lesser-known papers on gas theory and their relation to the same" (1879b) or even about a "hostile attitude" (1898a, v) towards gas theory, and of his awareness of "being a powerless individual struggling against the currents of the time".

Thus, the myth has arisen that Boltzmann was ignored or resisted by his contemporaries. Sometimes, his suicide in 1906 is attributed to the injustice he thus suffered, The fact that his death occurred just at the dawn of the definitive victory of the atomic view in the works of Einstein, Smoluchowski, Perrin et al. adds a further touch of drama to this picture.

As a matter of fact, Boltzmann's reputation as a theoretical physicist was actually widely known and well-respected. In 1888 he was offered (but declined, after a curious sequence of negotiations) a most prestigious chair in Berlin. Later, several universities (Vienna, Munich, Leipzig) competed to get him appointed, sometimes putting the salaries of several professorships together in their effort. He was elected to membership or honorary membership in many academies, received honorary doctorates, and was also awarded various medals. In short, there is no factual evidence for the claim that Boltzmann was ignored or suffered any unusual lack of recognition from his contemporaries.

Debates and Controversy

Boltzmann was involved in various disputes. But this is not to say that he was the innocent victim of hostilities. In many debates he took the initiative by launching a polemic attack on his colleagues. The most important debates were those with Mach and Ostwald on the reality of atoms, and with colleagues who criticised Boltzmann's own work in the form of the famous reversibility objection (Loschmidt) and the recurrence objection (Zermelo).

Ostwald and Mach clearly resisted the atomic view of matter (although for different reasons). Boltzmann certainly defended and promoted this view. But he was not the naive realist or unabashed believer in the existence of atoms that the more popular literature has made of him. Instead, he stressed from the 1880s onwards that the atomic view yielded at best an analogy, or a picture or model of reality. In his debate with Mach he advocated this approach as a useful or economical way to understand the thermal behaviour of gases. This means that his views were quite compatible with Mach's views on the goal of science.

What divided them was more a strategic issue. Boltzmann claimed that no approach in natural science that avoids hypotheses completely could ever succeed. He argued that those who reject the atomic hypothesis in favour of a continuum view of matter were guilty of adopting hypotheses too. Ultimately, the choice between such views should depend on their fruitfulness, and here Boltzmann had no doubt that the atomic hypothesis would be more successful. In the case of Ostwald, and his 'energetics', Boltzmann did become involved in a more heated dispute at a meeting in Lübeck in 1895. Roughly speaking, energetics presented a conception of nature that took energy as the most fundamental physical entity, and thus represented physical processes as transformations of various forms of energy. It resisted attempts to comprehend energy, or these transformations in terms of mechanical pictures.

It has been suggested that in the 1890s "the adherents of energetics reigned supreme in the German school and even throughout Europe". But this is surely a great exaggeration. It seems closer to the truth to say that energetics represented a rather small but vocal minority in the physics community, that claimed to put forward a seemingly attractive conception of natural science, and being promoted in the mid-90s by reputed scientists, could no longer be dismissed as the work of amateurs.

The 1895 gathering of the *Naturforscherversammlung* in Lübeck (the annual meeting of physicists, chemists, biologists and physicians) was programmed to devote special sessions to the state of the art of energetics. Boltzmann, who was member of the programme committee, had already shown interest in the development of energetics in private correspondence with Ostwald. Georg Helm was asked to prepare a report, and at Boltzmann's own suggestion, Ostwald also contributed a lecture. All agreed that the meeting should follow the "British style", i.e., manuscripts would be circulated beforehand and there would be ample room for discussion, following the example of the British Association for the Advancement of Science meeting that Boltzmann had attended the previous year.

Both Helm and Ostwald, apparently, anticipated that they would have the opportunity to discuss their views on energetics in an open-minded atmosphere. But at the meeting Boltzmann surprised them with devastating criticism. According to those who were present Boltzmann was the clear winner of the debate. Yet the energeticists experienced the confrontation as an ambush, for which he had not been prepared. Nevertheless, Boltzmann and Ostwald remained friends, and in 1902 Ostwald made a great effort to persuade his home university in Leipzig to appoint Boltzmann.

Neither is there any hostile attitude in the famous 'reversibility objection' by Loschmidt in 1875. Loschmidt was Boltzmann's former teacher and later colleague at the University of Vienna, and a life-long friend. He had no philosophical reservations against the existence of atoms at all. (Indeed, he is best known for his estimate of their size.) Rather, his main objection was against the prediction by Maxwell and Boltzmann that a gas column in thermal equilibrium in a gravitational field has the same temperature at all heights. His now famous reversibility objection arose in his attempts to undermine this prediction. Whether Boltzmann succeeded in refuting the objection or not is still a matter of dispute.

Zermelo's opposition had a quite different background. When he put forward the recurrence objection in 1896, he was an assistant to Planck in Berlin. And like his mentor, he did not favour the mechanical underpinning of thermal phenomena. Yet his 1896 paper is by no means hostile. It presents a careful logical argument that leads him to a dilemma: thermodynamics with its Second Law on the one hand and gas theory (in the form as Zermelo understood it) on the other cannot both be literally true. By contrast, it is Boltzmann's reaction to Zermelo, drenched in sarcasm and bitterness which (if anything) may have led to hostile feelings between these two authors. In any case, the tone of Zermelo's is considerably sharper. Still, Zermelo maintained a keen, yet critical, interest in gas theory and statistical physics, and subsequently played an important role in making Gibbs' work known in Germany.

In fact, I think that Boltzmann's rather aggressive reactions to Zermelo and Ostwald should be compared to other polemical exchanges in which he was involved, and sometimes initiated himself (e.g., against Clausius, Tait, Planck, and Bertrand — not to mention his essay on Schopenhauer). It seems to me that Boltzmann enjoyed polemics, and the use of sharp language for rhetorical effect. Boltzmann's complaints in 1896–1898 about an hostile environment are, I think, partly explained by his love of polemic exaggerations, partly also by his mental depression in that period. Certainly, the debates with Ostwald and Zermelo might well have contributed to this personal crisis. But it would be wrong to interpret Boltzmann's plaintive moods as evidence that his critics were, in fact, hostile.

Some praise them as brilliant and exceptionally clear. Often one finds passages suggesting he possessed all the right answers all along the way — or at least in his later writings, while his critics were simply prejudiced, confused or misguided (von Plato, Lebowitz, Kac, Bricmont, Goldstein). Others (Ehrenfests, Klein, Truesdell) have emphasised that Boltzmann's work is not always clear and that he often failed to indicate crucial assumptions or important changes in his position, while friendly critics helped him in clarifying and developing his views. Fans and critics of Boltzmann's work alike agree that he pioneered much of the approaches currently used in statistical physics, but also that he did not leave behind a unified coherent theory. His scientific papers, collected in *Wissenschaftliche Abhandlung,* contain more than 100 papers on statistical physics alone.

Some of these papers are forbiddingly long, full of tedious calculations and lack a clear coherent structure. Sometimes, vital assumptions, or even a complete change of approach, are stated only somewhere tucked away between the calculations, or at the very last page. Even Maxwell, who might have been in the best position to appreciate Boltzmann's work, expressed his difficulty with Boltzmann's long windedness. But not all of his prose is cumbersome and heavy-going. Boltzmann at his best could be witty, passionate and a delight to read. He excelled in such qualities in much of his popular work and some of his polemical articles.

Boltzmann's Relevance for the Basics of Statistical Physics

The foundations of statistical physics may today be characterised as a battlefield between a dozen or so different schools, each firmly dug into their own trenches, e.g., ergodic theory, coarse-graining (Markovianism), interventionism, BBKGY, Jaynes, Prigogine, etc. Still, many of the protagonists of these schools, regardless of their disagreements, frequently express their debt to ideas first formulated to Boltzmann. Even to those who consider the concept of ensembles as the most important tool of statistical physics, and claim Gibbs rather than Boltzmann as their champion, it has been pointed out that Boltzmann introduced ensembles long before Gibbs. And those who advocate Boltzmann while rejecting ergodic theory, may similarly be reminded that the latter theory too originated with Boltzmann himself.

It appears, therefore, that Boltzmann is the father of many approaches, even if these approaches are presently seen as conflicting with each other. This is due to the fact that during his forty years of work on the subject, Boltzmann pursued many lines of thought. Typically, he would follow a particular train of thought that seemed promising and fruitful, only to discard it in the next paper for another one, and then pick it up again years later. This meandering approach is of course not unusual among theoretical physicists but it makes it hard to pin down Boltzmann on a particular set of rock-bottom assumptions, that would reveal his true colours in the modern debate on the foundations of statistical physics. The Ehrenfests (1912) in their famous Encyclopaedia article, set themselves the task of constructing a more or less coherent framework out of Boltzmann's legacy. But their presentation of Boltzmann was, as is rather well known, not historically adequate.

Without going into a more detailed description of the landscape of the battlefield of the foundations of statistical physics, or a sketch of the various positions occupied, it might be useful to mention only the roughest of distinctions. I use the term 'statistical physics' as a deliberately vague term that includes at least two more sharply distinguished theories: the kinetic theory of gases and statistical mechanics proper.

The first theory aims to explain the properties of gases by assuming that they consist of a very large number of molecules in rapid motion. (The term 'kinetic' is meant to underline the vital importance of motion here, and to distinguish the approach from older static molecular gas models.) During the 1860s probability considerations were imported into this theory. The aim then became to characterise the properties of gases, in particular in thermal equilibrium, in terms of probabilities of various molecular states.

This is what the Ehrenfests call "kineto-statistics of the molecule". Here, molecular states, in particular their velocities, are regarded as stochastic variables, and probabilities are attached to such molecular states of motion. These probabilities themselves are conceived of as mechanical properties of the state of the total gas system. Either they represent the relative number of molecules with a particular state, or the relative time during which a molecule has that state. In the course of time a transition was made to what the Ehrenfests called "kineto-statistics of the gas model", or what is nowadays known as statistical mechanics. In this latter approach, probabilities are not attached to the state of a molecule but of the entire gas system. Thus, the state of the gas, instead of determining the probability distribution, now itself becomes a stochastic variable.

A merit of this latter approach is that interactions between molecules can be taken into account. Indeed, the approach is not restricted to gases, but also applicable to liquids or solids. The price to be paid, however, is that the probabilities themselves become more abstract. Since probabilities are attributed to the mechanical states of the total system, they are no longer determined by such mechanical states. Instead, in statistical mechanics, the probabilities are usually determined by means of an 'ensemble', i.e., a fictitious collection of replicas of the system in question.

It is not easy to pinpoint this transition in the course of history, except to say that in Maxwell's work in the 1860s definitely belong to the first category, and Gibbs' book of 1902 to the second. Boltzmann's own works fall somewhere in the middle. His earlier contributions clearly belong to the kinetic theory of gases (although his 1868 paper already applies probability to an entire gas system), while his work after 1877 is usually seen as elements in the theory of statistical mechanics. However, Boltzmann himself never indicated a clear distinction between these two different theories, and any attempt to draw a demarcation at an exact location in his work seems somewhat arbitrary. From a conceptual point of view, the transition from kinetic gas theory to statistical mechanics poses two main foundational questions. On what grounds do we choose a particular ensemble, or the probability distribution characterising the system? Gibbs did not enter into a systematic discussion of this problem, but only discussed special cases of equilibrium ensembles (i.e., canonical, micro-canonical etc.). A second problem is to relate the ensemble-based probabilities with the probabilities obtained in the earlier kinetic approach for a single gas model.

The Ehrenfests (1912) paper was the first to recognise these questions, and to provide a partial answer: Assuming a certain hypothesis of Boltzmann's, which they dubbed the ergodic hypothesis, they pointed out that for an isolated system the micro-canonical distribution is the unique stationary probability distribution. Hence, if one demands that an ensemble of isolated systems describing thermal equilibrium must be represented by a stationary distribution, the only choice for this purpose is the micro-canonical one. Similarly, they pointed out that under the ergodic hypothesis infinite time averages and ensemble averages were identical. This, then, would provide a desired link between the probabilities of the older kinetic gas theory and those of statistical mechanics, at least in equilibrium and in the infinite time limit. Yet the Ehrenfests simultaneously expressed strong doubts about the validity of the ergodic hypothesis. These doubts were soon substantiated when in 1913 Rozenthal and Plancherel proved that the hypothesis was untenable for realistic gas models.

The Ehrenfests' reconstruction of Boltzmann's work thus gave a prominent role to the ergodic hypothesis, suggesting that it played a fundamental and lasting role in his thinking. Although this view indeed produces a more coherent view of his multifaceted work, it is certainly not historically correct. Boltzmann himself also had grave doubts about this hypothesis, and expressly avoided it whenever he could, in particular in his two great papers of 1872 and 1877b. Since the Ehrenfests, many other authors have presented accounts of Boltzmann's work. Particularly important are Klein (1973) and Brush (1976). Still, much confusion remains about what exactly his approach to statistical physics was, and how it developed.

A Brief Chronography of Boltzmann's Writings

Roughly speaking, one may divide Boltzmann's work in four periods. The period 1866-1871 is more or less his formative period. In his first paper (1866), Boltzmann set himself the problem of deriving the full second law from mechanics. The notion of probability does not appear in this chapter. The following papers, from 1868 and 1871, were written after Boltzmann had read Maxwell's work of 1860 and 1867. Following Maxwell's example, they deal with the characterisation of a gas in thermal equilibrium, in terms of a probability distribution. Even then, he was set on obtaining more general results, and extended the discussion to cases where the gas is subject to a static external force, and might consist of poly-atomic molecules. He regularly switched between different conceptions of probability: sometimes this referred to a time average, sometimes a particle average or, in an exceptional paper (1871b), it referred to an ensemble average.

The main result of those papers is that from the Stosszahlansatz (SZA) (or an analogous assumption) the Maxwellian distribution function is stationary, and thus an appropriate candidate for the equilibrium state. In some cases Boltzmann also argued it was the unique such state. However, in this period he also presented a completely different method, which did not rely on the SZA but rather on the ergodic hypothesis. This approach led to a new form of the distribution function that, in the limit reduces to the Maxwellian form. In the same period, he also introduced the concept of ensembles, but this concept would not play a prominent role in his thinking until the 1880's.

The next period is that of 1872-1878, in which he wrote his two most famous papers: (1872) (Weitere Studien) and (1877b) (Über die Beziehung). The 1872 paper contained the Boltzmann equation and the H-theorem. Boltzmann claimed that the H-theorem provided the desired theorem from mechanics corresponding to the second law. However, this claim came under a serious objection due to Loschmidt's criticism of 1876. The objection was simply that no purely mechanical theorem could ever produce a time-asymmetrical result. Boltzmann's response to this objection will be summarised later.

The result was, however, that Boltzmann rethought the basis of his approach and in 1877b produced a conceptually very different analysis, which might be called the permutational argument, of equilibrium and evolutions towards equilibrium, and the role of probability theory. The distribution function, which formerly represented the probability distribution, was now conceived of as a stochastic variable (nowadays called a macrostate) subject to a probability distribution. That probability distribution was now determined by the size of the volume in phase space corresponding to all the microstates giving rise to the same macrostate, (essentially given by calculating all permutations of the particles in a given macrostate). Equilibrium was now conceived of as the most probable macrostate instead of a stationary macrostate. The evolution towards equilibrium could then be reformulated as an evolution from less probable to more probable states.

Even though all commentators agree on the importance of these two papers, there is still disagreement about what Boltzmann's claims actually were, and whether he succeeded (or indeed even attempted) in avoiding the reversibility objection in this new permutational argument, whether he intended or succeeded to prove that most evolutions go from less probable to more probable states and whether or not he (implicitly) relied on the ergodic hypothesis in these works. I shall comment on these issues in due course.

The third period is taken up by the papers Boltzmann wrote during the 1880's have attracted much less attention. During this period, he abandoned the permutational argument, and went back to an approach that relied on a combination of the ergodic hypothesis and the use of ensembles. For a while Boltzmann worked on an application of this approach to Helmholtz's concept of monocyclic systems. However, after finding that concept did not always provide the desired thermodynamical analogies, he abandoned this topic again. Next, in the 1890s the reversibility problem resurfaced again, this time in a debate in the columns of Nature.

This time Boltzmann chose an entirely different line of counterargument than in his debate with Loschmidt. A few years later, Zermelo presented another objection, now called the recurrence objection. The same period also saw the publication of the two volumes of his Lectures on Gas Theory. In this book, he takes the hypothesis of molecular disorder (a close relative of the SZA) as the basis of his approach.

The permutational argument is only discussed as an aside, and the ergodic hypothesis is not mentioned at all. His last paper is an Encyclopaedia article with Nabl presenting a survey of kinetic theory.

QUANTUM MECHANICS AND CLASSICAL PHYSICS

Predictions of quantum mechanics have been verified experimentally to an extremely high degree of accuracy. According to the correspondence principle between classical and quantum mechanics, all objects obey the laws of quantum mechanics, and classical mechanics is just an approximation for large systems of objects (or a statistical quantum mechanics of a large collection of particles).

The laws of classical mechanics thus follow from the laws of quantum mechanics as a statistical average at the limit of large systems or large quantum numbers. However, chaotic systems do not have good quantum numbers, and quantum chaos studies the relationship between classical and quantum descriptions in these systems.

Quantum coherence is an essential difference between classical and quantum theories, and is illustrated by the Einstein-Podolsky-Rosen paradox. Quantum interference involves adding together probability amplitudes, whereas classical "waves" infer that there is an adding together of intensities. For microscopic bodies, the extension of the system is much smaller than the coherence length, which gives rise to long-range entanglement and other nonlocal phenomena that are characteristic of quantum systems. Quantum coherence is not typically evident at macroscopic scales - although an exception to this rule can occur at extremely low temperatures (i.e. approaching absolute zero), when quantum behaviour can manifest itself on more macroscopic scales.

This is in accordance with the following observations:

- Many macroscopic properties of a classical system are a direct consequence of the quantum behaviour of its parts. For example, the stability of bulk matter (which consists of atoms and molecules which would quickly collapse under electric forces alone), the rigidity of solids, and the mechanical, thermal, chemical, optical and magnetic properties of matter are all results of the interaction of electric charges under the rules of quantum mechanics.
- While the seemingly "exotic" behaviour of matter posited by quantum mechanics and relativity theory become more apparent when dealing with particles of extremely small size or velocities approaching the speed of light, the laws of classical Newtonian physics remain accurate in predicting the behaviour of the vast majority of "large" objects (on the order of the size of large molecules or bigger) at velocities much smaller than the velocity of light.

❐

4

Interpretations and Success of Quantum Mechanics

Quantum mechanics is a physical science dealing with the behaviour of matter and energy on the scale of atoms and subatomic particles/ waves. Quantum mechanics is the foundation of several related disciplines, including condensed matter physics, quantum chemistry, and particle physics, and is the chief investigative tool in structural biology. Within the field of engineering, quantum mechanics plays an important part of forming the development of nanotechnology and electronics. The term "quantum mechanics" was first coined by Max Born in 1924. The acceptance by the general physics community of quantum mechanics is due to its accurate prediction of the physical behaviour of systems, including systems where Newtonian mechanics fails.

Even general relativity is limited—in ways quantum mechanics is not— in describing systems at the atomic scale or smaller, at very low or very high energies, or at the lowest temperatures. Through a century of experimentation and applied science, quantum mechanical theory has proven to be very successful and practical.

In advanced topics of quantum mechanics, some of these behaviours are macroscopic and only emerge at extreme (i.e., very low or very high) energies or temperatures. The name quantum mechanics derives from the observation that some physical quantities can change only in discrete amounts (Latin quanta), and not in a continuous (cf. analog) way. For example, the angular momentum of an electron bound to an atom or molecule is quantized. In the context of quantum mechanics, the wave–

particle duality of energy and matter and the uncertainty principle provide a unified view of the behaviour of photons, electrons, and other atomic-scale objects.

The mathematical formulations of quantum mechanics are abstract. A mathematical function called the wavefunction provides information about the probability amplitude of position, momentum, and other physical properties of a particle. Mathematical manipulations of the wavefunction usually involve the bra-ket notation, which requires an understanding of complex numbers and linear functionals. The wavefunction treats the object as a quantum harmonic oscillator, and the mathematics is akin to that describing acoustic resonance. Many of the results of quantum mechanics are not easily visualized in terms of classical mechanics - for instance, the ground state in a quantum mechanical model is a non-zero energy state that is the lowest permitted energy state of a system, as opposed a more "traditional" system that is thought of as simply being at rest, with zero kinetic energy. Instead of a traditional static, unchanging zero state, quantum mechanics allows for far more dynamic, chaotic possibilities, according to John Wheeler.

The earliest versions of quantum mechanics were formulated in the first decade of the 20th century. At around the same time, the atomic theory and the corpuscular theory of light (as updated by Einstein) first came to be widely accepted as scientific fact; these latter theories can be viewed as quantum theories of matter and electromagnetic radiation, respectively. Early quantum theory was significantly reformulated in the mid-1920s by Werner Heisenberg, Max Born, Wolfgang Pauli and their collaborators, and the Copenhagen interpretation of Niels Bohr became widely accepted. By 1930, quantum mechanics had been further unified and formalized by the work of Paul Dirac and John von Neumann, with a greater emphasis placed on measurement in quantum mechanics, the statistical nature of our knowledge of reality, and philosophical speculation about the role of the observer. Quantum mechanics has since branched out into almost every aspect of 20th century physics and other disciplines, such as quantum chemistry, quantum electronics, quantum optics, and quantum information science.

Much 19th century physics has been re-evaluated as the "classical limit" of quantum mechanics, and its more advanced developments in terms of quantum field theory, string theory, and speculative quantum gravity theories.

THE SURPRISING

Toward the end of the nineteenth century it seemed that physics had been very nearly perfected, but then paradox after paradox radically changed human concepts at fundamental levels. The models by which humans explained things on their scale in space and time failed to apply to the ultra small, the ultra massive, and to interactions at the highest speeds. A view of the universe that had seemed to be the distillation of commonplace observations was challenged by observations and theories that predicted correctly where classical physics had given impossible results. But the emerging picture was of a universe that stubbornly refused to behave the way humans had come to expect.

On the large end of the scale, relativity theory informed humans that time does not pass at the same rate for all observers, that matter can convert to energy and vice-versa, that two objects each moving at more than half the speed of light cannot have a closing speed that is equal to or greater than the speed of light, that time progresses more slowly near massive objects, etc. Things just do not work the way that experience with scales, rulers, and clocks here on earth had led humans to expect. On the small end of the scale, wonders are even more abundant.

A thing such as a photon or an electron does not have location and a traceable trajectory between the point where it is emitted and the point where it is detected. The points where such a particle may be detected are not what one would expect based on everyday experience. With a certain small probability, the detection point can even be on the other side of a solid barrier. Probability turns out to be a highly salient factor in all interactions on this scale.

The trajectory of any atomic scale object is "squishy" in the sense that any measurement intended to get clearer on the object's position forces knowledge of its velocity to become less clear, and vice-versa. In the era of classical physics, Newton and his followers believed that light consisted of particles, and others believed that light was actually a series of wave fronts propagated through some medium.

Rather than finding an experiment that would prove one or the other side correct, physicists learned that making an experiment designed to show light's frequency or other wave characteristics will manifest its "wave nature," and making an experiment that is designed to show its momentum or other particle characteristics will manifest its "particle nature." Furthermore, objects as large as atoms, and even some molecules, have shown their "wave nature" when addressed in the appropriate way. The most eminent of physicists have warned that if one imagines some explanation of quantum physics to "make sense," then that explanation is very likely to be flawed.

HOW THE UNFORESEEN CAME TO LIGHT

The foundations of quantum mechanics date from the early 1800s, but the real beginnings of QM date from the work of Max Planck in 1900. Albert Einstein and Niels Bohr soon made important contributions to what is now called the "old quantum theory."

However, it was not until 1924 that a more complete picture emerged with Louis de Broglie's matter-wave hypothesis and the true importance of quantum mechanics became clear. Some of the most prominent scientists to subsequently contribute in the mid-1920s to what is now called the "new quantum mechanics" or "new physics" were Max Born, Paul Dirac, Werner Heisenberg, Wolfgang Pauli, and Erwin Schrödinger.

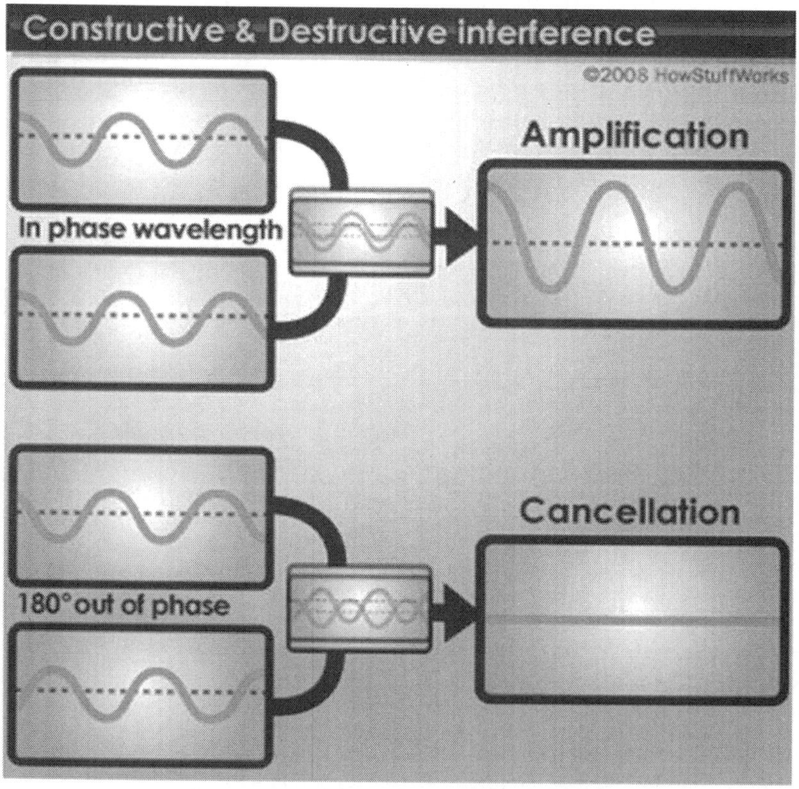

Fig.4.1: Constructive and Distructive Interfence

Later, the field was further expanded with work by Julian Schwinger, Sin-Itiro Tomonaga and Richard Feynman for the development of Quantum Electrodynamics in 1947 and by Murray Gell-Mann in particular for the development of Quantum chromodynamics. The interference that produces coloured bands on bubbles cannot be explained by a model that depicts light as a particle. It can be explained by a model that depicts it as a wave. The drawing shows sine waves that resemble waves on the surface of water being reflected from two surfaces of a film of varying width, but that depiction of the wave nature of light is only a crude analogy. Early researchers differed in their explanations of the fundamental nature of what we now call electromagnetic radiation.

Some maintained that light and other frequencies of electromagnetic radiation are composed of particles, while others asserted that electromagnetic radiation is a wave phenomenon. In classical physics these ideas are mutually contradictory. Ever since the early days of QM, scientists have acknowledged that neither idea by itself can explain electromagnetic radiation. In 1690, Christiaan Huygens explained the laws of reflection and refraction on the basis of a wave theory. Sir Isaac Newton believed that light consisted of infinitesimally small particles which he designated "corpuscles".

In 1827 Thomas Young and Augustin Fresnel made experiments on interference that showed that a corpuscular theory of light was inadequate. Then in 1873 James Clerk Maxwell showed that by making an electrical circuit oscillate it should be possible to produce electromagnetic waves. His theory made it possible to compute the speed of electromagnetic radiation purely on the basis of electrical and magnetic measurements, and the computed value corresponded very closely to the empirically measured speed of light. In 1888, Heinrich Hertz made an electrical device that actually produced what we would now call microwaves — essentially radiation at a lower frequency than visible light.

Everything up to that point suggested that Newton had been entirely wrong to regard light as corpuscular. Later experiments indicated that a packet or quantum model was needed to explain some phenomena. When light strikes an electrical conductor it causes electrons to move away from their original positions. The observed phenomenon could only be explained by assuming that the light delivers energy in definite packets. In a photoelectric device such as the light meter in a camera, light hitting the metallic detector causes electrons to move. Greater intensities of light at one frequency can cause more electrons to move, but they will not move faster. In contrast, higher frequencies of light can cause electrons to move faster.

Therefore, intensity of light controls current, but frequency of light controls voltage. These observations raised a contradiction when compared with sound waves and ocean waves, where only intensity was needed to predict the energy of the wave.

In the case of light, frequency appeared to predict energy. Something was needed to explain this phenomenon and to reconcile experiments that had shown light to have particle nature with experiments that had shown it to have wave nature.

SPECTROSCOPY AND FORWARD

It is fairly easy to see a spectrum produced by white light when it passes through a prism, the bevelled edge of a mirror or a tapered pane of glass, or through drops of rain to form a rainbow. When samples which are composed purely of one single chemical element, e.g. hydrogen, are caused to emit light, they may emit light at several characteristic frequencies. The frequency profile produced is characteristic of that element.

Instead of there being a wide band filled with colours from violet to red, there will be isolated bands of single colours separated by darkness. Such a display is called a line spectrum. Some lines go beyond the visible frequencies and can only be detected by special photographic film or other such devices.

Scientists hypothesized that an atom could radiate light the way the string on a fine violin radiates sound – not only with a fundamental frequency but with several higher harmonics (formed when the string divides itself into halves and other divisions that vibrate in coordination with each other as when one half of the string is going one way as the other half of the string is going the opposite way). For a long time nobody could find a mathematical way to relate the frequencies of the line spectrum of any element.

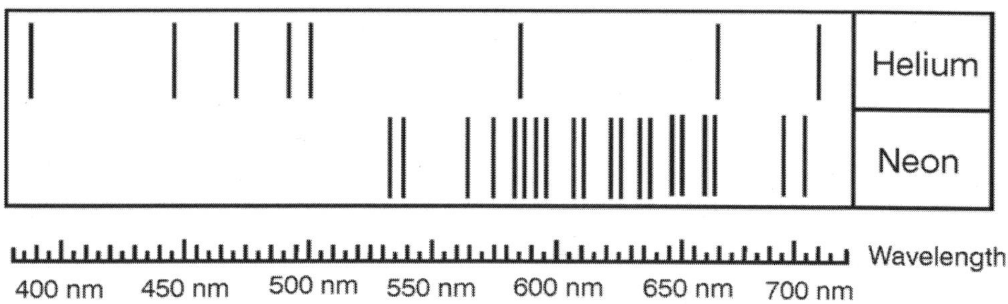

Fig.4.2: Bright-line Spectrum of Hydrogen

In 1885, Johann Jakob Balmer figured out how the frequencies of atomic hydrogen are related to each other. The formula is a simple one:

$$\frac{1}{\lambda} = R\left(\frac{1}{2^2} - \frac{1}{n^2}\right)$$

where λ is wavelength, R is the Rydberg constant and n is an integer (n =3, 4...) This formula can be generalized to apply to atoms that are more complicated than hydrogen, but we will stay with hydrogen for this general exposition. The next development was the discovery of the Zeeman effect, named after Pieter Zeeman.

The physical explanation of the Zeeman effect was worked out by Hendrik Antoon Lorentz. Lorentz hypothesized that the light emitted by hydrogen was produced by vibrating electrons. It was possible to get feedback on what goes on within the atom because moving electrons create a magnetic field and so can be influenced by the imposition of an external magnetic field in a manner analogous to the way that one iron magnet will attract or repel another magnet. The Zeeman effect could be interpreted to mean that light waves are originated by electrons vibrating in their orbits, but classical physics could not explain why electrons should not fall out of their orbits and into the nucleus of their atoms, nor could classical physics explain why their orbits would be such as to produce the series of discrete frequencies derived by Balmer's formula and displayed in the line spectra. Why did the electrons not produce a continuous spectrum?

INTERPRETATIONS AND SUCCESS OF QUANTUM MECHANICS

Despite the success of quantum mechanics, it does have some controversial elements. For example, the behaviour of microscopic objects described in quantum mechanics is very different from our everyday experience, which may provoke some degree of incredulity. Most of classical physics is now recognized to be composed of special cases of quantum physics theory and/ or relativity theory. Dirac brought relativity theory to bear on quantum physics so that it could properly deal with events that occur at a substantial fraction of the speed of light. Classical physics, however, also deals with mass attraction (gravity), and no one has yet been able to bring gravity into a unified theory with the relativized quantum theory. Also interesting, according to the correspondence principle and Ehrenfest's theorem as a system becomes larger or more massive (action >> h) the classical dynamics tends to emerge (with some exceptions such as superfluidity).

This is why we can usually ignore quantum mechanics when dealing with everyday objects; instead the classical description will suffice. Even so, trying to make sense of quantum theory is an ongoing process that has spawned a number of interpretations of quantum theory, ranging from the conventional Copenhagen Interpretation to hidden variables and many worlds interpretations. There seems to be no end in sight to the philosophical musings on the subject; however the empirical or technical success of the theory is unrivaled; all modern fundamental physical theories are quantum theories, relativity being subsumed within quantum field theories.

ATTEMPTS AT A UNIFIED ASSUMPTION

Inconsistencies arise when one tries to join the quantum laws with general relativity, a more elaborate description of spacetime which incorporates gravitation. Resolving these inconsistencies has been a major goal of twentieth- and twenty-first-century physics. Many prominent physicists, including Stephen Hawking, have labored in the attempt to discover a "Grand Unification Theory" that combines not only different models of subatomic physics, but also derives the universe's four forces— the strong force, electromagnetism, weak force, and gravity— from a single force or phenomenon.

GENERAL RELATIVITY AND QUANTUM MECHANICS

The modern world of physics is founded on two tested and demonstrably sound theories of general relativity and quantum mechanics —theories which appear to contradict one another. The defining postulates of both Einstein's theory of relativity and quantum theory are indisputably supported by rigorous and repeated empirical evidence. However, while they do not directly contradict each other theoretically they are resistant to being incorporated within one cohesive model. Einstein himself is well known for rejecting some of the claims of quantum mechanics. While clearly inventive in this field, he did not accept the more philosophical consequences and interpretations of quantum mechanics, such as the lack of deterministic causality and the assertion that a single subatomic particle can occupy numerous areas of space at one time. He also was the first to notice some of the apparently exotic consequences of entanglement and used them to formulate the Einstein- Podolsky-Rosen paradox, in the hope of showing that quantum mechanics has unacceptable implications.

This was 1935, but in 1964 it was shown by John Bell that Einstein's assumption that quantum mechanics is correct, but has to be completed by hidden variables, was based on wrong philosophical assumptions: according to the paper of J. Bell and the Copenhagen interpretation (the common interpretation of quantum mechanics by physicists for decades), and contrary to Einstein's ideas, quantum mechanics is,

- Neither a "realistic" theory (since quantum measurements do not state pre-existing properties, but rather they prepare properties)
- Nor a local theory (essentially not, because the state vector $|\psi>$ determines simultaneously the probability amplitudes at all sites, $|\psi>$.

The Einstein-Podolsky-Rosen paradox shows in any case that there exist experiments by which one can measure the state of one particle and instantaneously change the state of its entangled partner, although the two particles can be an arbitrary distance apart; however, this effect does not violate causality, since no transfer of information happens. These experiments are the basis of some of the most topical applications of the theory, quantum cryptography, which works well, although at small distances of typically ≤ 1000 km,

being on the market since 2004. There do exist quantum theories which incorporate special relativity-for example, quantum electrodynamics (QED), which is currently the most accurately tested physical theory-and these lie at the very heart of modern particle physics. Gravity is negligible in many areas of particle physics, so that unification between general relativity and quantum mechanics is not an urgent issue in those applications. However, the lack of a correct theory of quantum gravity is an important issue in cosmology.

PREDICTIONS OF QUANTUM MECHANICS

Predictions of quantum mechanics have been verified experimentally to a very high degree of accuracy. Thus, the current logic of correspondence principle between classical and quantum mechanics is that all objects obey laws of quantum mechanics, and classical mechanics is just a quantum mechanics of large systems (or a statistical quantum mechanics of a large collection of particles). Laws of classical mechanics thus follow from laws of quantum mechanics at the limit of large systems or large quantum numbers. Main differences between classical and quantum theories have already in the remarks on the Einstein-Podolsky-Rosen paradox.

Essentially the difference boils down to the statement that quantum mechanics is coherent (addition of amplitudes), whereas classical theories are incoherent (addition of intensities). Thus, such quantities as coherence lengths and coherence times come into play. For microscopic bodies the extension of the system is certainly much smaller than the coherence length; for macroscopic bodies one expects that it should be the other way round. Many "macroscopic" properties of "classic" systems are direct consequences of quantum behaviour of its parts. For example, stability of bulk matter (which consists of atoms and molecules which would quickly collapse under electric forces alone), rigidity of this matter, mechanical, thermal, chemical, optical and magnetic properties of this matter—they are all results of interaction of electric charges under the rules of quantum mechanics.

Whilst seemingly exotic behaviour of matter posited by quantum mechanics and relativity theory become more apparent when dealing with extremely fast-moving or extremely tiny particles, the laws of classical "Newtonian" physics still remain accurate in predicting the behaviour of surrounding ("large") objects—of the order of the size of large molecules and bigger—at velocities much smaller than the velocity of light.

There are numerous mathematically equivalent formulations of quantum mechanics. One of the oldest and most commonly used formulations is the transformation theory proposed by Cambridge theoretical physicist Paul Dirac, which unifies and generalizes the two earliest formulations of quantum mechanics, matrix mechanics (invented by Werner Heisenberg) and wave mechanics. In this formulation, the instantaneous state of a quantum system encodes the probabilities of its measurable properties, or "observables". Examples of observables include energy, position, momentum, and angular momentum. Observables can be either continuous (e.g., the position of a particle) or discrete (e.g., the energy of an electron bound to a hydrogen atom). Generally, quantum mechanics does not assign definite values to observables.

Instead, it makes predictions about probability distributions; that is, the probability of obtaining each of the possible outcomes from measuring an observable. Naturally, these probabilities will depend on the quantum state at the instant of the measurement. There are, however, certain states that are associated with a definite value of a particular observable. These are known as "eigenstates" of the observable ("eigen" can be roughly translated from German as inherent or as a characteristic). In the everyday world, it is natural and intuitive to think of everything being in an eigenstate of every observable.

Everything appears to have a definite position, a definite momentum, and a definite time of occurrence. However, quantum mechanics does not pinpoint the exact values for the position or momentum of a certain particle in a given space in a finite time; rather, it only provides a range of probabilities of where that particle might be. Therefore, it became necessary to use different words for the state of something having an uncertainty relation and a state that has a definite value. The latter is called the "eigenstate" of the property being measured.

For example, consider a free particle. In quantum mechanics, there is wave- particle duality so the properties of the particle can be described as a wave. Therefore, its quantum state can be represented as a wave, of arbitrary shape and extending over all of space, called a wave function. The position and momentum of the particle are observables. The Uncertainty Principle of quantum mechanics states that both the position and the momentum cannot simultaneously be known with infinite precision at the same time. However, one can measure just the position alone of a moving free particle creating an eigenstate of position with a wavefunction that is very large at a particular position x, and almost zero everywhere else. If one performs a position measurement on such a wavefunction, the result x will be obtained with almost 100% probability. In other words, the position of the free particle will almost be known. This is called an eigenstate of position (mathematically more precise: a generalized eigenstate (eigendistribution)). If the particle is in an eigenstate of position then its momentum is completely unknown.

It can be shown that the wavelength is equal to h/p, where h is Planck's constant and p is the momentum of the eigenstate. If the particle is in an eigenstate of momentum then its position is completely blurred out.

Usually, a system will not be in an eigenstate of whatever observable we are interested in. However, if one measures the observable, the wavefunction will instantaneously be an eigenstate of that observable. This process is known as wavefunction collapse. It involves expanding the system under study to include the measurement device, so that a detailed quantum calculation would no longer be feasible and a classical description must be used. If one knows the corresponding wave function at the instant before the measurement, one will be able to compute the probability of collapsing into each of the possible eigenstates.

For example, the free particle in the previous example will usually have a wavefunction that is a wave packet centered around some mean position x0, neither an eigenstate of position nor of momentum. When one measures the position of the particle, it is impossible to predict with certainty the result that we will obtain. It is probable, but not certain, that it will be near x0, where the amplitude of the wave function is large. After the measurement is performed, having obtained some result x, the wave function collapses into a position eigenstate centered at x.

Wave functions can change as time progresses. An equation known as the Schrödinger equation describes how wave functions change in time, a role similar to Newton's second law in classical mechanics. The Schrödinger equation, applied to the aforementioned example of the free particle, predicts that the centre of a wave packet will move through space at a constant velocity, like a classical particle with no forces acting on it.

However, the wave packet will also spread out as time progresses, which means that the position becomes more uncertain. This also has the effect of turning position eigenstates into broadened wave packets that are no longer position eigenstates. Some wave functions produce probability distributions that are constant in time. Many systems that are treated dynamically in classical mechanics are described by such "static" wave functions. For example, a single electron in an unexcited atom is pictured classically as a particle moving in a circular trajectory around the atomic nucleus, whereas in quantum mechanics it is described by a static, spherically symmetric wavefunction surrounding the nucleus. The time evolution of wave functions is deterministic in the sense that, given a wavefunction at an initial time, it makes a definite prediction of what the wavefunction will be at any later time.

During a measurement, the change of the wavefunction into another one is not deterministic, but rather unpredictable, i.e., random. The probabilistic nature of quantum mechanics thus stems from the act of measurement. This is one of the most difficult aspects of quantum systems to understand.

It was the central topic in the famous Bohr-Einstein debates, in which the two scientists attempted to clarify these fundamental principles by way of thought experiments. In the decades after the formulation of quantum mechanics, the question of what constitutes a "measurement" has been extensively studied. Interpretations of quantum mechanics have been formulated to do away with the concept of "wavefunction collapse" the relative state interpretation. The basic idea is that when a quantum system interacts with a measuring apparatus, their respective wavefunctions become entangled, so that the original quantum system ceases to exist as an independent entity.

Fundamental Rules of Quantum Mechanics

The fundamental rules of quantum mechanics are very broad. They assert that the state space of a system is a Hilbert space and the observables are Hermitian operators acting on that space, but do not tell us which Hilbert space or which operators, or if it even exists. These must be chosen appropriately in order to obtain a quantitative description of a quantum system. An important guide for making these choices is the correspondence principle, which states that the predictions of quantum mechanics reduce to those of classical physics when a system moves to higher energies or equivalently, larger quantum numbers.

In other words, classic mechanics is simply a quantum mechanics of large systems. This "high energy" limit is known as the classical or correspondence limit. One can therefore start from an established classical model of a particular system, and attempt to guess the underlying quantum model that gives rise to the classical model in the correspondence limit.

When quantum mechanics was originally formulated, it was applied to models whose correspondence limit was non-relativistic classical mechanics. For instance, the well-known model of the quantum harmonic oscillator uses an explicitly non-relativistic expression for the kinetic energy of the oscillator, and is thus a quantum version of the classical harmonic oscillator. Early attempts to merge quantum mechanics with special relativity involved the replacement of the Schrödinger equation with a covariant equation such as the Klein-Gordon equation or the Dirac equation. While these theories were successful in explaining many experimental results, they had certain unsatisfactory qualities stemming from their neglect of the relativistic creation and annihilation of particles. A fully relativistic quantum theory required the development of quantum field theory, which applies quantization to a field rather than a fixed set of particles. The first complete quantum field theory, quantum electrodynamics, provides a fully quantum description of the electromagnetic interaction.

The full apparatus of quantum field theory is often unnecessary for describing electrodynamic systems. A simpler approach, one employed since the inception of quantum mechanics, is to treat charged particles as quantum mechanical objects being acted on by a classical electromagnetic field. For example, the elementary quantum model of the hydrogen atom describes the electric field of the hydrogen atom using a classical, Coulomb potential. This "semi-classical" approach fails if quantum fluctuations in the electromagnetic field play an important role, such as in the emission of photons by charged particles. Quantum field theories for the strong nuclear force and the weak nuclear force have been developed. The quantum field theory of the strong nuclear force is called quantum chromodynamics, and describes the interactions of the subnuclear particles: quarks and gluons.

The weak nuclear force and the electromagnetic force were unified, in their quantized forms, into a single quantum field theory known as electroweak theory, by the physicists Abdus Salam, Sheldon Glashow and Steven Weinberg. It has proven difficult to construct quantum models of gravity, the remaining fundamental force.

Semi-classical approximations are workable, and have led to predictions such as Hawking radiation. However, the formulation of a complete theory of quantum gravity is hindered by apparent incompatibilities between general relativity, the most accurate theory of gravity currently known, and some of the fundamental assumptions of quantum theory.

The resolution of these incompatibilities is an area of active research, and theories such as string theory are among the possible candidates for a future theory of quantum gravity.

Mathematical Formulation of Quantum Mechanics

In the mathematically rigorous formulation of quantum mechanics, developed by Paul Dirac and John von Neumann, the possible states of a quantum mechanical system are represented by unit vectors (called "state vectors") residing in a complex separable Hilbert space well defined up to a complex number of norm 1 (the phase factor). In other words, the possible states are points in the projectivization of a Hilbert space, usually called the complex projective space. The exact nature of this Hilbert space is dependent on the system; for example, the state space for position and momentum states is the space of square-integrable functions, while the state space for the spin of a single proton is just the product of two complex planes.

Each observable is represented by a maximally-Hermitian linear operator acting on the state space. Each eigenstate of an observable corresponds to an eigenvector of the operator, and the associated eigenvalue corresponds to the value of the observable in that eigenstate. If the operator's spectrum is discrete, the observable can only attain those discrete eigenvalues. The time evolution of a quantum state is described by the Schrödinger equation, in which the Hamiltonian, the operator corresponding to the total energy of the system, generates time evolution.

The inner product between two state vectors is a complex number known as a probability amplitude. During a measurement, the probability that a system collapses from a given initial state to a particular eigenstate is given by the square of the absolute value of the probability amplitudes between the initial and final states. The possible results of a measurement are the eigenvalues of the operator - which explains the choice of Hermitian operators, for which all the eigenvalues are real. We can find the probability distribution of an observable in a given state by computing the spectral decomposition of the corresponding operator. Heisenberg's uncertainty principle is represented by the statement that the operators corresponding to certain observables do not commute. The Schrödinger equation acts on the entire probability amplitude, not merely its absolute value. Whereas the absolute value of the probability amplitude encodes information about probabilities, its phase encodes information about the interference between quantum states. This gives rise to the wave-like behaviour of quantum states.

It turns out that analytic solutions of Schrödinger's equation are only available for a small number of model Hamiltonians, of which the quantum harmonic oscillator, the particle in a box, the hydrogen molecular ion and the hydrogen atom are the most important representatives. Even the helium atom, which contains just one more electron than hydrogen, defies all attempts at a fully analytic treatment. There exist several techniques for generating approximate solutions. For instance, in the method known as perturbation theory one uses the analytic results for a simple quantum mechanical model to generate results for a more complicated model related to the simple model by, for example, the addition of a weak potential energy.

Another method is the "semi-classical equation of motion" approach, which applies to systems for which quantum mechanics produces weak deviations from classical behaviour. The deviations can be calculated based on the classical motion. This approach is important for the field of quantum chaos. An alternative formulation of quantum mechanics is Feynman's path integral formulation, in which a quantum-mechanical amplitude is considered as a sum over histories between initial and final states; this is the quantum- mechanical counterpart of action principles in classical mechanics.

SOURCE OF QUANTIZATION

The particle in a 1-dimensional potential energy box is the most simple example where restraints lead to the quantization of energy levels. The box is defined as zero potential energy inside a certain interval and infinite everywhere outside that interval. For the 1-dimensional case in the x direction, the time-independent Schrödinger equation can be written as:

$$\frac{\hbar}{2m}\frac{d^2\Psi}{dx^2} = \psi$$

The general solutions are:

$$y = Ae^{ikx} + Be-^{ikx} \quad E = \frac{k^2\hbar^2}{2m}$$

$$y = C\sin kx + D\cos kx \text{ (exponential rewrite)}$$

The presence of the walls of the box restricts the acceptable solutions to the wavefunction. At each wall:

$$\psi = 0 \; at \; x = 0, x = L$$

Consider x = 0

- Sin 0 = 0, cos 0 = 1. To satisfy y = 0 D = 0 (cos term is removed) Now Consider: y = C sin kx
- At X = L, y = C sin kL
- If C = 0 then for all x and would conflict with Born interpretation
- Therefore sin kL must be satisfied by

In this situation, n must be an integer showing the quantization of the energy levels.

Applications of Quantum Mechanics

Quantum mechanics has had enormous success in explaining many of the features of our world. The individual behaviour of the subatomic particles that make up all forms of matter—electrons, protons, neutrons, photons and others—can often only be satisfactorily described using quantum mechanics. Quantum mechanics has strongly influenced string theory, a candidate for a theory of everything and the multiverse hypothesis. It is also related to statistical mechanics. Quantum mechanics is important for understanding how individual atoms combine covalently to form chemicals or molecules. The application of quantum mechanics to chemistry is known as quantum chemistry. Quantum mechanics can in principle mathematically describe most of chemistry. Quantum mechanics can provide quantitative insight into ionic and covalent bonding processes by explicitly showing which molecules are energetically favourable to which others, and by approximately how much. Most of the calculations performed in computational chemistry rely on quantum mechanics. Much of modern technology operates at a scale where quantum effects are significant. Examples include the laser, the transistor, the electron microscope, and magnetic resonance imaging. The study of semiconductors led to the invention of the diode and the transistor, which are indispensable for modern electronics. Researchers are currently seeking robust methods of directly manipulating quantum states.

Efforts are being made to develop quantum cryptography, which will allow guaranteed secure transmission of information. A more distant goal is the development of quantum computers, which are expected to perform certain computational tasks exponentially faster than classical computers.

Another active research topic is quantum teleportation, which deals with techniques to transmit quantum states over arbitrary distances. In many devices, even the simple light switch, quantum tunneling is vital, as otherwise the electrons in the electric current could not penetrate the potential barrier made up, in the case of the light switch, of a layer of oxide. Flash memory chips found in USB drives also use quantum tunneling to erase their memory cells.

PHILOSOPHICAL CONSEQUENCES OF QUANTUM MECHANICS

Since its inception, the many counter-intuitive results of quantum mechanics have provoked strong philosophical debate and many interpretations. Even fundamental issues such as Max Born's basic rules concerning probability amplitudes and probability distributions took decades to be appreciated.

The Copenhagen interpretation, due largely to the Danish theoretical physicist Niels Bohr, is the interpretation of quantum mechanics most widely accepted amongst physicists. According to it, the probabilistic nature of quantum mechanics predictions cannot be explained in terms of some other deterministic theory, and does not simply reflect our limited knowledge. Quantum mechanics provides probabilistic results because the physical universe is itself probabilistic rather than deterministic. Albert Einstein, himself one of the founders of quantum theory, disliked this loss of determinism in measurement. He held that there should be a local hidden variable theory underlying quantum mechanics and consequently the present theory was incomplete. He produced a series of objections to the theory, the most famous of which has become known as the EPR paradox. John Bell showed that the EPR paradox led to experimentally testable differences between quantum mechanics and local realistic theories. Experiments have been performed confirming the accuracy of quantum mechanics, thus demonstrating that the physical world cannot be described by local realistic theories. The writer C. S. Lewis viewed quantum mechanics as incomplete.

Lewis, a professor of English, was of the opinion that the Heisenberg uncertainty principle was more of an epistemic limitation than an indication of ontological indeterminacy, and in this respect believed similarly to many advocates of hidden variables theories.

The Bohr-Einstein debates provide a vibrant critique of the Copenhagen Interpretation from an epistemological point of view. The Everett many-worlds interpretation, formulated in 1956, holds that all the possibilities described by quantum theory simultaneously occur in a "multiverse" composed of mostly independent parallel universes.

This is not accomplished by introducing some new axiom to quantum mechanics, but on the contrary by removing the axiom of the collapse of the wave packet: All the possible consistent states of the measured system and the measuring apparatus are present in a real physical quantum superposition. While the multiverse is deterministic, we perceive non-deterministic behaviour governed by probabilities, because we can observe only the universe, i.e. the consistent state contribution to the mentioned superposition, we inhabit.

Everett's interpretation is perfectly consistent with John Bell's experiments and makes them intuitively understandable. However, according to the theory of quantum decoherence, the parallel universes will never be accessible to us.

This inaccessibility can be understood as follows: once a measurement is done, the measured system becomes entangled with both the physicist who measured it and a huge number of other particles, some of which are photons flying away towards the other end of the universe; in order to prove that the wave function did not collapse one would have to bring all these particles back and measure them again, together with the system that was measured originally. This is completely impractical, but even if one can theoretically do this, it would destroy any evidence that the original measurement took place.

QUANTUM MECHANICAL THEORY

Werner Heisenberg made the full quantum mechanical theory in 1925 at the young age of 23. Following his mentor, Niels Bohr, Werner Heisenberg began to work out a theory for the quantum behaviour of electron orbitals. Because electrons could not be observed in their orbits, Heisenberg went about creating a mathematical description of quantum mechanics built on what could be observed, that is, the light emitted from atoms in their characteristic atomic spectra. Heisenberg studied the electron orbital on the model of a charged ball on a spring, an oscillator, whose motion is anharmonic.

Heisenberg first explained this kind of observed motion in terms of the laws of classical mechanics known to apply in the macro world, and then applied quantum restrictions, discrete (non-continuous) properties, to the picture. Doing so causes gaps to appear between the predicted orbitals so that the mathematical description he formulated would then represent only the electron orbitals predicted on the basis of the atomic spectra. In approaching the problem that Bohr gave him to solve, Heisenberg took the strategic stance that he would not deal with unobservable quantities. He would begin formulating equations using only the quantities that could be observed. That strategy led him to begin with the actual experimental evidence at hand.

Measurements had been well established for such data as:

- The frequencies (and the mathematically related energies) emitted or absorbed by electron transitions from one of the Bohr stationary orbits, known to be associated with the bright line spectra.
- The "transition amplitude" or likelihood of transition from any given orbit to any given orbit, known from the strength of the various lines in the bright spectrum, etc.

From classical formulas that would characterize those phenomena Heisenberg created analogous formulas that took account of quantum conditions. Formulas that followed from the fundamental decisions made at this point resulted in good results but results that were sometimes not what one might expect. In the paper wherein he introduced quantum mechanics to the world he cautions, "A significant difficulty arises, however, if we consider two quantities x(t), y(t), and ask after their product. Whereas in classical x(t)y(t) is always equal to y(t)x(t), this is not necessarily the case in quantum theory." When the predicted values are exhibited in matrix form and multiplications are performed, the nature of the difficulty appears in a form that is more familiar to mathematicians. More significantly, empirical studies validate the theoretical results and suggest that there is something of deep importance in that the difference between x(t)y(t) and y(t)x(t) is a value related to Planck's constant.

Schema for a table of transition frequencies (produced when electrons change orbitals):

Electron States	**S1**	**S2**	**S3**	**S4**	**S5**	**....**
S1	f1→1	f2→1	f3→1	f4→1	f5→1	.....
S2	f1→2	f2→2	f3→2	f4→2	f5→2	.....
S3	f1→3	f2→3	f3→3	f4→3	f5→3	.....
S4	f1→4	f2→4	f3→4	f4→4	f5→4	.....
S5	f1→5	f2→5	f3→5	f4→5	f5→5	.....
S.....	.....	.....	.....	.....	.....	.....

Schema for a related table showing the transition amplitudes:

Electron States	**S1**	**S2**	**S3**	**S4**	**S5**	**....**
S1	a1′→1	a2′→1	a3′→1	a4′→1	a5′→1	.....
S2	a1′→2	a2′→2	a3′→2	a4′→2	a5′→2	.....
S3	a1′→3	a2′→3	a3′→3	a4′→3	a5′→3	.....
S4	a1′→4	a2′→4	a3′→4	a4′→4	a5′→4	.....
S5	a1′→5	a2′→5	a3′→5	a4′→5	a5′→5	.....
S.....	.....	.....	.....	.....	.....	.....

Heisenberg developed ways of meaningfully relating the information in tables such as these in a mathematical way. Empirically filling in the values for tables involving quantum quantities is not a simple procedure since any measurement made on a single system gives that one value but has the potential of changing other values. So large numbers of identical copies of the system in question must be prepared, and a single measure made on each system. Multiple experiments to determine the same characteristics are made, and the results are averaged. Even then, precise measurements of all characteristics of the system as they would appear simultaneously cannot be provided because of quantum uncertainty. A precise determination of one characteristic's value necessarily creates an uncertainty in the value of its correlate. "Certain pairs of observables simply cannot be simultaneously measured to an arbitrarily high level of precision. " If simultaneous measurements are made of correlated characteristics in multiple identical systems, there will inevitably be differences in the measurements such that the difference between their products is equal to or greater than h/2."

In 1925 Heisenberg published a paper entitled "Quantum-mechanical re- interpretation of kinematic and mechanical relations" relating his discoveries. So ended the old quantum theory and began the age of quantum mechanics. Heisenberg's paper gave few details that might aid readers in determining how he actually contrived to get his results for the one-dimensional models he used to form the hypothesis that proved so useful.

In his paper, Heisenberg proposed to "discard all hope of observing hitherto unobservable quantities, such as the position and period of the electron", and restrict himself strictly to actually observable quantities. He needed mathematical rules for predicting the relations actually observed in nature, and the rules he produced worked differently depending on the sequence in which they were applied. "It quickly became clear that the non- commutativity of kinematical quantities in quantum theory was the really essential new technical idea in the paper."

The special type of multiplication that turned out to be required in his formula was most elegantly described by using special arrays of numbers called matrices. In ordinary situations it does not matter in which order the operations involved in multiplication are performed, but matrix multiplication does not commute. Essentially that means that it matters which order given operations are performed in. Multiplying matrix A by matrix B is not the same as multiplying matrix B by matrix A.

In symbols, $A \times B$ is in general not equal to $B \times A$. (The important thing in quantum theory is that it turned out to matter whether one experimentally measures velocity first and then immediately measures position, or vice-versa.) The matrix convention turned out to be a convenient way of organizing information and making clear the exact sequence in which calculations must be made, and it reflects in a symbolic form the unexpected results obtained in the real world.

Heisenberg approached quantum mechanics from the historical perspective that treated an electron as an oscillating charged particle. Bohr's use of this analogy had already allowed him to explain why the radii of the orbits of electrons could only take on certain values. It followed from this interpretation of the experimental results available and the quantum theory that Heisenberg subsequently created that an electron could not be at any intermediate position between two "permitted" orbits. Therefore electrons were described as "jumping" from orbit to orbit.

The idea that an electron might now be in one place and an instant later be in some other place without having traveled between the two points was one of the earliest indications of the "spookiness" of quantum phenomena. Although the scale is smaller, the "jump" from orbit to orbit is as strange and unexpected as would be a case in which someone stepped out of a doorway in London onto the streets of Los Angeles.

Quantum tunnelling is one instance in which electrons seem to be able to move in the "spooky" way that Heisenberg ascribed to their actions within atoms. Amplitudes of position and momentum that have a period of 2 p like a cycle in a wave are called Fourier series variables. Heisenberg described the particle-like properties of the electron in a wave as having position and momentum in his matrix mechanics. When these amplitudes of position and momentum are measured and multiplied together, they give intensity. However, he found that when the position and momentum were multiplied together in that respective order, and then the momentum and position were multiplied together in that respective order, there was a difference or deviation in intensity between them of h/2p.

Heisenberg would not understand the reason for this deviation until two more years had passed, but for the time being he satisfied himself with the idea that the math worked and provided an exact description of the quantum behaviour of the electron. Matrix mechanics was the first complete definition of quantum mechanics, its laws, and properties that described fully the behaviour of the electron. It was later extended to apply to all subatomic particles.

Very soon after matrix mechanics was introduced to the world, Schrödinger, acting independently, produced a quantum wave theory that appeared to have no similarities whatsoever to Heisenberg's theory. It was computationally easier and avoided some of the odd-sounding ideas like "quantum leaps" of an electron from one orbit to another. But within a short time Schrödinger himself had shown that the two theories produced essentially the same results in all situations. Finally, Dirac made the idea of non- commutativity central to his own theory and proved the formulations of Heisenberg and of Schrödinger to be special cases of his own theory.

Schrodinger Wave Equation

Because particles could be described as waves, later in 1925 Erwin Schrödinger analysed what an electron would look like as a wave around the nucleus of the atom. Using this model, he formulated his equation for particle waves. Rather than explaining the atom by analogy to satellites in planetary orbits, he treated everything as waves whereby each electron has its own unique wavefunction. A wavefunction is described in Schrödinger's equation by three properties.

The three properties were:

- An "orbital" designation, indicating whether the particle wave is one that is closer to the nucleus with less energy or one that is farther from the nucleus with more energy,
- The shape of the orbital, i.e., an indication that orbitals were not just spherical but other shapes, and
- The inclination of the orbital, which determines the magnetic moment of the orbital around the z-axis.

These three properties were called collectively the wavefunction of the electron and are said to describe the quantum state of the electron. "Quantum state" means the collective properties of the electron describing what we can say about its condition at a given time. For the electron, the quantum state is described by its wavefunction, which is designated in physics by the Greek letter ø (psi, pronounced "sigh"). The three properties of Schrödinger's equation that describe the wavefunction of the electron and, therefore, also describe the quantum state of the electron as described in the previous paragraph are each called quantum numbers. The first property that describes the orbital was numbered according to Bohr's model where n is the letter used to describe the energy of each orbital. This is called the principal quantum number.

The next quantum number that describes the shape of the orbital is called the azimuthal quantum number and it is represented by the letter l (lower case L). The shape is caused by the angular momentum of the orbital. The rate of change of the angular momentum of any system is equal to the resultant external torque acting on that system. In other words, angular momentum represents the resistance of a spinning object to speed up or slow down under the influence of external force. The azimuthal quantum number "l" represents the orbital angular momentum of an electron around its nucleus. However, the shape of each orbital has its own letter as well. So for the letter "l" there are other letters to describe the shapes of "l".

The next shape is like a dumbbell and is described by the letter p. The other shapes of orbitals become more complicated and are described by the letters d, f, and g. For the shape of a carbon atom. The third quantum number of Schrödinger's equation describes the magnetic moment of the electron and is designated by the letter m and sometimes as the letter m with a subscript l because the magnetic moment depends upon the second quantum number l. In May 1926 Schrödinger published a proof that Heisenberg's matrix mechanics and his own wave mechanics gave equivalent results: mathematically they were the same theory. Yet both men disagreed on the interpretation of this theory. Heisenberg saw no problem in the existence of discontinuous quantum jumps, while Schrödinger hoped that a theory based on continuous wave-like properties could avoid this "nonsense about quantum jumps".

Indecision Principle

In 1927, Heisenberg made a new discovery on the basis of his quantum theory that had further practical consequences of this new way of looking at matter and energy on the atomic scale. In Heisenberg's matrix mechanics formula, he encountered an error or difference of $h/2\pi$ between position and momentum.

The more certain the position of a particle is determined, the less certain the momentum is known, and $h/2\pi$ is the lower limit of the uncertainty involved. This conclusion came to be called "Heisenberg's Indeterminacy Principle", or Heisenberg's Uncertainty Principle. For moving particles in quantum mechanics, there is simply a certain degree of exactness and precision that is missing. The observer can be precise when taking a measurement of position or can be precise when taking a measurement of momentum, but there is an inverse imprecision when measuring both at the same time as in the case of a moving particle like the electron. One consequence of the uncertainty principle was that the electron could no longer be considered as in an exact location in its orbital.

Rather the electron had to be described by every point where the electron could possibly inhabit. Calculating points of probable location for the electron in its known orbital created the picture of a cloud of points in a spherical shape for the orbital of a hydrogen atom which points gradually faded out nearer to the nucleus and farther from the nucleus. This picture may be termed a probability distribution. Therefore, the Bohr atom number n for each orbital became known as an n-sphere in the three dimensional atom and was pictured as a probability cloud where the electron surrounded the atom all at once. This led to the further description by Heisenberg that if a measurement of the electron was not being taken that it could not be described in one particular location but was everywhere in the electron cloud at once. In other words, quantum mechanics cannot give exact results, but only the probabilities for the occurrence of a variety of possible results.

Heisenberg went further and said that the path of a moving particle only comes into existence once we observe it. However strange and counter-intuitive this assertion may seem, quantum mechanics does still tell us the location of the electron's orbital, its probability cloud. Heisenberg was speaking of the particle itself, not its orbital which is in a known probability distribution. It is important to note that although Heisenberg used infinite sets of positions for the electron in his matrices, this does not mean that the electron could be anywhere in the universe. Rather there are several laws that show the electron must be in one localized probability distribution. An electron is described by its energy in Bohr's atom which was carried over to matrix mechanics.

Therefore, an electron in a certain n-sphere had to be within a certain range from the nucleus depending upon its energy. This restricts its location. Also, the number of places an electron can be is also called "the number of cells in its phase space".

The Uncertainty Principle set a lower limit to how finely one can chop up classical phase space, so the number of places that an electron can be in its orbital becomes finite. An electron's location in an atom is defined to be in its orbital, but stops at the nucleus and before the next n-sphere orbital begins. Classical physics had shown since Newton that if the position of stars and planets and details about their motions were known then where they will be in the future can be predicted. For subatomic particles, Heisenberg denied this notion showing that due to the uncertainty principle one cannot know the precise position and momentum of a particle at a given instant, so its future motion cannot be determined, but only a range of possibilities for the future motion of the particle can be described.

These notions arising from the uncertainty principle only arise at the subatomic level and were a consequence of wave-particle duality. As counter- intuitive as they may seem, quantum mechanical theory with its uncertainty principle has been responsible for major improvements in the world's technology from computer components to fluorescent lights to brain scanning techniques.

THE PAULI EXCLUSION PRINCIPLE

The Pauli exclusion principle is the quantum mechanical principle that no two identical fermions (particles with half-integer spin) may occupy the same quantum state simultaneously. A more rigorous statement is that the total wave function for two identical fermions is anti-symmetric with respect to exchange of the particles. The principle was formulated by Austrian physicist Wolfgang Pauli in 1925. The Pauli Exclusion Principle states that no electron (or other fermion) can be in the same quantum state as another within an atom. Wolfgang Pauli developed the Exclusion Principle from what he called a "two-valued quantum degree of freedom" to account for the observation of a doublet, meaning a pair of lines differing by a small amount (e.g., something on the order of 0.15Å), in the spectrum of the hydrogen atom.

The observation meant that there was more energy in the electron orbital from magnetic moment than had previously been described. In early 1925, the young physicists Uhlenbeck and Goudsmit introduced a theory that the electron rotates in space in the same way that the earth rotates on its axis. This would account for the missing magnetic moment and allow for two electrons in the same orbital to be different if their spins were in opposite directions, thus satisfying the Exclusion Principle. A new quantum number was then needed, one to represent the momentum involved in the rotation of each electron.

One concise statement of Pauli's principle is as follows: There cannot exist an atom in such a quantum state that two electrons within have the same set of quantum numbers.

An electron has four quantum numbers:

- *n*, the total quantum number of the orbit
- *l*, the azimuthal quantum number
- *ml*, the orbital magnetic quantum number
- *ms*, the spin magnetic quantum number

"In the case of a helium atom with two electrons in the 1s orbital, the Pauli exclusion principle requires that the two electrons differ in the value of one quantum number. Their values of n, l, and ml are the same; moreover, they have the same spin quantum number, s = ½. Accordingly they must differ in the value of ms, which can have the value of +½ for one electron and –½ for the other."

Quantum Embarrassment

Albert Einstein rejected Heisenberg's Uncertainty Principle insofar as it seemed to imply more than a necessary limitation on human ability to actually know what occurs in the quantum realm. In a letter to Max Born in 1926, Einstein famously declared that "God does not play dice". The bare surface level prescription for making predictions from quantum mechanics, based on Born's rule for computing probabilities, became known as the Copenhagen Interpretation of quantum mechanics.

Bohr spent many years developing and refining this interpretation in light of Einstein's objections. After the 1930 Solvay conference, Einstein never again challenged the Copenhagen interpretation on technical grounds, but did not cease a philosophical attack on the interpretation, on the grounds of realism and locality (i.e., the idea that there is no action at a distance). Einstein, in trying to show that quantum theory was not a complete theory, recognized that the theory predicted that two or more particles that had interacted in the past can exhibit strong correlations later on when various measurements are made on them. He wanted this seeming interaction to be explained in a classical way through their common past, and preferably not by a "spooky action at a distance". The argument is worked out in a famous paper devoted to what is now called the EPR paradox. Assuming what is now usually called "local realism", the

EPR attempts to show from quantum theory that particles simultaneously possess both position and momentum, while according to the Copenhagen interpretation, only one of these two properties actually exists, and only in the moment that it is being measured.

Einstein considered this conclusion a proof that quantum theory is incomplete in that it refuses to physical properties which objectively exist in nature. The feature of quantum theory leading to these paradoxes is called quantum entanglement. Quantum entanglement refers to situations wherein the properties of several discrete objects cannot be described simply by considering them separately, even after taking account of the history of their past interaction.

A quantum interaction can affect the entangled quantum entity in a like manner even though the two of them are widely separated. Bohr's original response to Einstein was that entangled particles are part of one indivisible system.

Einstein's challenge led to decades of substantial research into quantum entanglement. Research would seem to confirm Bohr's assertion that two or more entangled particles must be viewed together as one whole, and, moreover, that difficulties only arise by insisting on the reality of outcomes of measurements that are not made anyway.

It might be said that God does indeed throw dice, though rather peculiar ones. A real dice throw can be completely understood using the laws of classical mechanics, and the outcome is merely a function of the initial conditions. However the outcome of tossing quantum "dice" has no antecedent; it has as yet no discernible cause.

Dirac Wave Equation

In 1928, Paul Dirac extended the Pauli equation, which described spinning electrons, to account for relativity. By using the simplest electromagnetic interaction, he was able to predict the value of the magnetic moment associated with the electron's spin, and found the experimentally observed value, a value too large to be that of a spinning classical charged sphere. He was able to solve for the spectral lines of the Hydrogen atom, reproducing Sommerfeld's successful formula for the fine-structure splitting of the spectral lines from sound physical principles.

Dirac's theory suffered from a problem of negative energy, which required Dirac to introduce antiparticles and a dynamical vacuum and led to many- particle quantum field theory. The development of quantum theory was based mainly on the atomic spectrum of the hydrogen atom. This is due to the fact that each atom of each element produces a unique pattern of spectral lines when light from each different kind of element is passed through a prism. Scientists could not study the electron and nucleus of the atom itself because they cannot be seen.

Even today with High-resolution Scanning Tunneling Electron Microscopes we can only get images of the atom as a blurry fuzzball. However, the spectral lines of the atom reveal the orbits of electrons and the energy that can be expected. It was basically this study of the spectroscopic analysis of first the hydrogen atom and then the helium atom that led to quantum theory. Therefore, the mathematical formulas were made to fit the picture of the atomic spectrum. That is why quantum mechanics is sometimes referred to as a form of mathematical physics.

WAVE FUNCTION COLLAPSE

In quantum mechanics, wave function collapse is the phenomenon in which a wave function—initially in a superposition of several different possible eigenstates—appears to reduce to a single one of those states after interaction with an observer. In simplified terms, it is the reduction of the physical possibilities into a single possibility as seen by an observer. It is one of two processes by which quantum systems evolve in time, according to the laws of quantum mechanics as presented by John von Neumann. The reality of wave function collapse has always been debated, i.e., whether it is a fundamental physical phenomenon in its own right or just an epiphenomenon of another process, such as quantum decoherence. In recent decades the quantum decoherence view has gained popularity and is commonly taught at the graduate level. Collapse may be understood as an update in a probabilistic model, given the observed result.

Schrödinger's wave equation with its unique wavefunction for a single electron is also spread out in a probability distribution like Heisenberg's quantized particle-like electron. This is because a wave is naturally a widespread disturbance and not a point particle.

Therefore, Schrödinger's wave equation has the same predictions made by the uncertainty principle because uncertainty of location is built into the definition of a widespread disturbance like a wave. Uncertainty only needed to be defined from Heisenberg's matrix mechanics because the treatment was from the particle-like aspects of the electron. Schrödinger's wave equation shows that the electron is in the probability cloud at all times in its probability distribution as a wave that is spread out. Max Born discovered in 1928 that when the square of Schrödinger's wavefunction (psi-squared) is computed the electron's location as a probability distribution is obtained.

Therefore, to the extent that a measurement of the position of an electron can be made at an exact location instead of as a probability distribution, the electron appears to momentarily cease to have wave-like properties. Without wave-like properties, none of Schrödinger's definitions of the electron being wave-like makes sense. The measurement of the position of the particle nullifies the simple wave-like properties and the one body form of Schrödinger's equation then fails.

Because the electron can no longer be described by its separate wavefunction when measured, due to its wave length becoming much shorter and its becoming entangled with the particles of the measuring apparatus, this is called wavcfunction collapse.

Eigenstates and Eigenvalues

The term eigenstate is derived from the German/Dutch word "eigen," which means "inherent" or "characteristic." The word eigenstate is descriptive of the measured state of some object that possesses quantifiable characteristics such as position, momentum, etc. The state being measured and described must be an "observable" (i.e. something that can be experimentally measured either directly or indirectly like position or momentum), and must have a definite value. In the everyday world, it is natural and intuitive to think of everything being in its own eigenstate.

Everything appears to have a definite position, a definite momentum, a definite value of measure, and a definite time of occurrence. However, quantum mechanics affirms that it is impossible to pinpoint exact values for the momentum of a certain particle like an electron in a given location at a particular moment in time, or, alternatively, that it is impossible to give an exact location for such an object when the momentum has been measured. Due to the uncertainty principle, statements regarding both the position and momentum of particles can only be given in terms of a range of probabilities, a "probability distribution". Eliminating uncertainty in one term maximizes uncertainty in regard to the second parameter.

Therefore it became necessary to have a way to clearly formulate the difference between the state of something that is uncertain in the way just described, such as an electron in a probability cloud, and effectively contrast it to the state of something that is not uncertain, something that has a definite value. When something is in the condition of being definitely "pinned-down" in some regard, it is said to possess an eigenstate. For example, if the position of an electron has been made definite, it is said to have an eigenstate of position. A definite value, such as the position of an electron that has been successfully located, is called the eigenvalue of the eigenstate of position. The German word "eigen" was first used in this context by the mathematician David Hilbert in 1904.

Schrödinger's wave equation gives wavefunction solutions, meaning the possibilities where the electron might be, just as does Heisenberg's probability distribution. When a wavefunction collapse occurs because something has been done to locate the position of an electron, the electron's state becomes an eigenstate of position, meaning that the position has a known value.

THE QUANTUM OF ANGULAR MOMENTUM

In physics, angular momentum, moment of momentum, or rotational momentum is a vector quantity that can be used to describe the overall state of a physical system. However, Bohr found a clever theoretical way to determine RH from his model: by equating his prediction of the frequency emitted when an electron goes from one orbit to another in a very large atom with the classical prediction which would be just the orbital frequency of the electron, how many times per second it goes around, he deduced,

$$K = 1/2\pi$$

and from that the Rydberg constant that appeared before is here given in terms of h, m and e. The rather abstract argument that the quantum predictions must match the known classical results for large slow systems actually fixes the Rydberg constant.

His argument goes as follows: for the circular orbits:

$$\frac{mv^2}{r} = \frac{1}{4\pi\varepsilon_0} \bullet \frac{e^2}{r^2} \text{ so } mv^2 = \frac{1}{4\pi\varepsilon_0} \bullet \frac{e^2}{r^2}, K.E. = -\frac{1}{2}P.E, E = -\frac{1}{4\pi\varepsilon_0} \bullet \frac{e^2}{r}$$

With the angular momentum quantized, for the nth orbit:

$$mv_n r_n = nKh,$$

where h is Planck's constant, *n* an integer, *K* the unknown multiplying factor (ok, fixed by experiment, but we're finding it independently). From this quantization condition we can find the radius, and hence the energy, of the nth orbit:

$$\frac{1}{4\pi\varepsilon_0} \bullet \frac{e^2}{r_n} = mv_n^2 = m\left(\frac{nkh}{mr_n}\right)^2$$

Giving,

$$r_n = \frac{4\pi\varepsilon_0 n^2 k^2 h^2}{me^2}, E_n = \left(\frac{1}{4\pi\varepsilon_0}\right) \bullet \frac{me^4}{2K^2h^2} \bullet \frac{1}{n^2}$$

In the large n limit,

$$E_{n+1} - E \cong \left(\frac{1}{4\pi\varepsilon_0}\right)^2 \bullet \frac{me^4}{2K^2h^2} \bullet \frac{2}{n^3} = h\nu$$

so,

$$\nu = \left(\frac{1}{4\pi\varepsilon_0}\right)^2 \bullet \frac{me^4}{K^2h^2n^3}$$

where v is the frequency of the emitted photon on jumping down one quantum number.

In the classical limit of large n, v must match the orbital frequency of the electron, since Maxwell's equations will be valid.

This formula was found to be correct within the limits of experimental error in measuring the quantities on the right. But few people believed his theory. For one thing, it soon became apparent that in the spectra of some stars (actually including the sun) there were spectral lines apparently corresponding to half the angular momentum quantum.

How could that be? Bohr's response was that these lines must be from ionized helium, not hydrogen. A neutral helium atom has two electrons, a singly-ionized helium atom has just one electron, but the nucleus has a charge twice that of the hydrogen nucleus, so the factor e4 in the Rydberg constant is replaced by 4e4, which leads to the observed result and found that actually the RH for these new lines corresponded to a factor of 4.0016. How could Bohr explain that? Bohr pointed out that at this level of precision, the finite mass of the nucleus must be taken into account by using a reduced mass for the electron. This gives just the right factor. This result greatly impressed Einstein, who concluded that Bohr must be on the right track.

THE BOHR ATOM IS STILL IMPORTANT

Bohr's semiclassical analysis has long been replaced by Schrödinger's wave function, there are recent experiments in atomic physics where the classical approach provides valuable insight. In particular, so-called Rydberg atoms, which are atoms with one electron in a spatially large orbit (large n, weakly bound), act a lot like classical systems. Such atoms can be ionized by microwave fields. For a considerable range of parameters, the onset of this ionization can be accounted for by ignoring quantum mechanics altogether, and interpreting ionization as the onset of chaotic motion in the classical driven system!

And, the standard perturbation theoretic methods of quantum mechanics don't work for this system anyway, because the perturbing microwave electric field is of the same order of magnitude as the atom's electric field at these large orbits. We should mention that, counterintuitively, quantum mechanics does become important again at very large n (or high microwave frequency), where some tricks from condensed matter physics have been used successfully to interpret the experiments. This is a rich subject: qualitatively different phenomena occur as the ratio of microwave frequency to orbital frequency is varied.

THE PHONE COMPANY MAKES EVERYTHING CRYSTAL OBVIOUS

There was an accident at the Bell Telephone Laboratories in April 1925. Clinton Davisson and L. H. Germer, looking for ways to improve vacuum tubes, were watching how electrons from an electron gun in a vacuum tube scattered off a flat nickel surface. Suddenly, while the experiment was running and the nickel target was very hot, a bottle of liquid air near the apparatus exploded, smashing one of the vacuum pipes, and air rushed into the apparatus.

The hot nickel target oxidized immediately. The layer of oxide made their target useless for further investigations. They decided to clean off the oxide by heating the nickel in a hydrogen atmosphere then in vacuum. After doing this for a prolonged period, the nickel looked good, and they resumed the investigation.

To their amazement, the pattern of electron scattering from the newly cleaned nickel target was completely different from that before the accident. What had changed? On examining their newly cleaned crystal carefully, they found a clue. The original target was polycrystalline made up of a multitude of tiny crystals, oriented randomly. During the prolonged heating of the cleaning process, the nickel had re-crystallized into a few large crystals. To quote from their paper: "It seemed probable to us from these results that the intensity of scattering from a single crystal would exhibit a marked dependence on crystal direction, and we set about at once preparing experiments for an investigation of this dependence. We must admit that the results obtained in these experiments have proved to be quite at variance with our expectations.

It seemed likely that strong beams would be found issuing from the crystal along what may be termed its transparent directions the directions in which the atoms in the lattice are arranged along the smallest number of lines per unit area. Strong beams are indeed found issuing from the crystal, but only when the speed of bombardment lies near one or another of a series of critical values, and then in directions quite unrelated to crystal transparency.

"The most striking characteristic of these beams is a one to one correspondence...which the strongest of them bear to the Laue beams that would be found issuing from the same crystal if the incident beam were a beam of x-rays. Certain others appear to be analogues of optical diffraction beams from plane reflection gratings the lines of these gratings being lines or rows of atoms in the surface of the crystal. Because of these similarities a description in terms of an equivalent wave radiation is not only possible, but most simple and natural.

This involves the association of a wavelength with the incident electron beam, and this wavelength turns out to be in acceptable agreement with the value h/mv of the undulatory mechanics, Planck's action constant divided by the momentum of the electron. "That evidence for the wave nature of particle mechanics would be found in the reaction between a beam of electrons and a single crystal was predicted by Elsasser two years ago shortly after the appearance of L. de Broglie's original papers on wave mechanics." It should be added that the two-slit diffraction pattern is of course exhibited by a beam of electrons, has been observed experimentally many times, and has precisely the same form as that for light.

MAXWELL'S WAVE EQUATION

Let us examine what Maxwell's equations tell us about the motion of the simplest type of electromagnetic wave a monochromatic wave in empty space, with no currents or charges present.

WAVE EQUATIONS AND WAVEPACKETS

De Broglie's doctoral thesis, defended at the end of 1924, created a lot of excitement in European physics circles. Shortly after it was published in the fall of 1925 Pieter Debye, a theorist in Zurich, suggested to Erwin Schrodinger that he give a seminar on de Broglie's work. Schrodinger gave a polished presentation, but at the end Debye remarked that he considered the whole theory rather childish: why should a wave confine itself to a circle in space?

It wasn't as if the circle was a waving circular string, real waves in space diffracted and diffused, in fact they obeyed three-dimensional wave equations, and that was what was needed. This was a direct challenge to Schrodinger, who spent some weeks in the Swiss mountains working on the problem, and constructing his equation.

There is no rigorous derivation of Schrodinger's equation from previously established theory, but it can be made very plausible by thinking about the connection between light waves and photons, and construction an analogous structure for de Broglie's waves and electrons.

THE WAVE EQUATION TELL US ABOUT THE PHOTON

We know from the photoelectric effect and Compton scattering that the photon energy and momentum are related to the frequency and wavelength of the light by,

$$E = h\nu = \hbar\omega$$

$$p = \frac{h}{\lambda} = \hbar k$$

Notice, then, that the wave equation tells us that $\omega = ck$ and hence $E = cp$.

To put it another way, if we think of a particle (photon) it would be more natural to write the plane wave as, that is, in terms of the energy and momentum of the particle.

In these terms, applying the (Maxwell) wave equation operator to the plane wave yields,

$$\left(\frac{\partial^2}{\partial x^2} - \frac{1}{c^2}\frac{\partial^2}{\partial t^2}\right)\vec{E}_0 e^{\frac{i}{\hbar}(px-Et)} = \left(p^2 - \frac{E^2}{c^2}\right)\vec{E}_0 e^{\frac{i}{\hbar}(px-Et)} = 0$$

or,

$$E^2 = c^2p^2$$

The wave equation operator applied to the plane wave describing the particle propagation yields the energy-momentum relationship for the particle.

CONSTRUCTING A WAVE EQUATION OPERATOR

The wave equation operator from the photon case (zero rest mass) to a particle having rest mass m0. We need a wave equation operator that, when it operates on a plane wave, yields,

$$E^2 = c^2 p^2 + m^2 c^4$$

Writing the plane wave function,

$$\phi(x,t) = Ae^{\frac{i}{\hbar}(pk-Et)}$$

where A is a constant, we find we can get $E^2 = c^2 p^2 + m^2 c^4$ by adding a constant (mass) term to the differentiation terms in the wave operator:

$$\left(\frac{\partial^2}{\partial x^2} - \frac{1}{c^2}\frac{\partial^2}{\partial t^2}\right)\vec{E}_0 e^{\frac{i}{\hbar}(px-Et)} = \left(p^2 - \frac{E^2}{c^2}\right)\vec{E}_0 e^{\frac{i}{\hbar}(px-Et)} = 0$$

This wave equation is called the Klein-Gordon equation and correctly describes the propagation of relativistic particles of mass m0. However, it's a bit inconvenient for nonrelativistic particles, like the electron in the hydrogen atom, just as E2 = m02c4 + c2p2 is less useful than E= p2/2m for this case.

A NONRELATIVISTIC WAVE EQUATION

Continuing along the same lines, let us assume that a nonrelativistic electron in free space (no potentials, so no forces) is described by a plane wave.

We need to construct a wave equation operator which, applied to this wave function, just gives us the ordinary nonrelativistic energy-momentum relationship, E = p2/2m. The p2 obviously comes as usual from differentiating twice with respect to x, but the only way we can get E is by having a single differentiation with respect to time, so this looks different from previous wave equations

This is Schrodinger's equation for a free particle. It is easy to check that it has the plane wave form given above, the condition for it to be a solution of this wave equation is just $E = p^2/2m$.

Notice one remarkable feature of the above equation the i on the left means that ψ cannot be a real function.

VARYING POTENTIAL AFFECT A DE BROGLIE WAVE

The effect of a potential on a de Broglie wave was considered by Sommerfeld in an attempt to generalize the rather restrictive conditions in Bohr's model of the atom. Since the electron was orbiting in an inverse square force, just like the planets around the sun, Sommerfeld couldn't understand why Bohr's atom had only circular orbits, no Kepler-like ellipses. (Recall that all the observed spectral lines of hydrogen were accounted for by energy differences between circular orbits.) De Broglie's analysis of the allowed circular orbits can be formulated by assuming at some instant in time the spatial variation of the wave function on going around the orbit includes a phase term of the form eipq /ψ , where here the parameter q measures distance around the orbit.

Now for an acceptable wave function, the total phase change on going around the orbit must be 2nψ , where n is an integer. For the usual Bohr circular orbit, p is constant on going around, q changes by 2ψr, where r is the radius of the orbit, the usual angular momentum quantization. What Sommerfeld did was to consider a general Kepler ellipse orbit, and visualize the wave going around such an orbit. Assuming the usual relationship vary as the particle moves around the orbit, being shortest where the particle moves fastest, at its closest approach to the nucleus. Nevertheless, the phase change on moving a short distance ψq should still be pψq/ψ, and requiring the wave function to link up smoothly on going once around the orbit..

Thus only certain elliptical orbits are allowed. The mathematics is nontrivial, but it turns out that every allowed elliptical orbit has the same energy as one of the allowed circular orbits.

That is why Bohr's theory gave all the energy levels. Actually, this whole analysis is old fashioned (it's called the "old quantum theory") but we've gone over it to introduce the idea of a wave with variable wavelength, changing with the momentum as the particle moves through a varying potential. The reader may well be wondering at this point why it is at all useful to visualize a real wave going round an orbit, when we have stated that any solution of Schrodinger's equation is necessarily a complex function. Including those corresponding to Bohr's energy levels, in which the complex nature of the wave function only appears in a time varying phase factor eψiEt/ψ . We should also add that if the spatial dependence is a real function, such as sin kx, it represents a standing wave, not a particle circling in one direction, which would be eikx, or e Bearing all this in mind, it is still often instructive to sketch real wave functions, especially for one-dimensional problems.

SCHRODINGER'S EQUATION FOR A PARTICLE IN A POTENTIAL

In quantum mechanics and scattering theory, the one dimensional step potential is an idealized system used to model incident, reflected and transmitted matter waves. The problem consists of solving the time-independent Schrödinger equation for a particle with a step-like potential in one dimension. Typically, the potential is modelled as a Heaviside step function.

Let us consider first the one-dimensional situation of a particle going in the x-direction subject to a "roller coaster" potential. What do we expect the wave function to look like?

We would expect the wavelength to be shortest where the potential is lowest, in the valleys, because that's where the particle is going fastest maximum momentum. With a nonzero potential present, the energy-momentum relationship for the particle becomes the energy equation.

We need to construct a wave equation which leads naturally to this relationship. In contrast to the free particle cases, the relevant wave function here will no longer be a plane wave, since the wavelength varies with the potential. However, at a given x, the momentum is determined by the "local wavelength".

This is the standard one-dimensional Schrodinger equation.

In three dimensions, the argument is precisely analogous. The only difference is that the square of the momentum is now a sum of three squared components.

This is the complete Schrodinger equation. So far, of course, it is based on plausibility arguments and hand-waving. Why should anyone believe that it really describes an electron wave? Schrodinger's test of his equation was the hydrogen atom. He looked for Bohr's "stationary states": states in which the electron was localized somewhere near the proton, and having a definite energy. The time dependence would be the same as for a plane wave of definite energy, eψiEx/ψ the spatial dependence would be a time-independent function decreasing rapidly at large distances from the proton. That is, he took, advantage of the spherical symmetry by re-expressing the spatial wave function in spherical polar coordinates, and found his equation became a standard differential equation solved in the nineteenth century. The solution gave the shape of possible wave functions, and also allowed values of energy and angular momentum.

These values were exactly the same as Bohr's (except that the lowest allowed state in the new theory had zero angular momentum): impressive evidence that the new theory was correct.

PRESENT CONSERVATION

When Schrodinger published this result in 1926, he also wrote down the complex conjugate equation, and proved that taking them together it was not difficult to deduce a continuity equation.

Schrodinger believed the above continuity equations represented the conservation of electric charge, and had no further significance. He thought that after all his own equation showed the electron to be just a smooth classical wave at the deepest level. In fact, he succeeded in solving the three-dimensional equation with a Coulomb potential and he found the Bohr energy levels of the hydrogen atom.

Obviously, he was on the right track! This classical revival approach, however, couldn't deal with the unpredictability of quantum mechanics, such as where a single photon or electron would land in a two-slit diffraction pattern. The truth is, Schrodinger didn't understand his own equation.

Another physicist, Max Born, published a paper a few days after Schrodinger's in which he suggested that, was the relative probability of finding the electron in a small volume *dxdydz* at (x,y,z) at time t. This interpretation was based directly on the analogy with light waves and photons, and has turned out to be correct. Notation note is called the *"amplitude"* or sometimes the *"probability amplitude"*.

QUANTUM EQUILIBRIUM

Perception

Equilibrium is the state of an object when no part of the object is accelerating. Two conditions for equilibrium are that the net force acting on the object is zero, and the net torque acting on the object is zero. Architectural structures are normally stationary. Most clients, building officials and designers prefer that their structures remain static rather than move dynamically. There are specific loading conditions which are dynamic loads, but in each and every case a return to a stable and static state is desireable. Such a condition is known as equilibrium.

Various states of static equilibrium are experienced throughout one's life. Think of the "teeter-totter" at a playground or of a game of "tug-of-war." In the first case, two or more individuals sit upon a board which has been fixed to a fulcrum which allows rotation. If each of the individuals on the teeter totter weight exactly the same amount and sit at exactly the same distance from the fulcrum the teeter-totter will not move. A state of equilibrium has been achieved. The two will remain at rest until an action takes them out of equilibrium. Such an action could be the addition of another person to the system or it could be that one of the original two would change their position slightly. In either case, the teeter-totter would most likely swing to one side and rest upon the ground. A new state of equilibrium would have been found. In this case, another boy climbed on behind the one already sitting on the right. In order to put the system back into equilibrium the girl on the left had to move backwards along the board until she was far away. The moment two boys created around the fulcrum doubled when the second boy climbed on. The girl knew that the only way that she could increase the magnitude of the moment she created would be to increase the moment arm.

A structure is in equilibrium when all forces or moments acting upon it are balanced. This means that each and every force acting upon a body, or part of the body, is resisted by either another equal and opposite force or set of forces whose net result is zero. Issac Newton addressed this issue when he noted that a body is at rest will remain at rest until acted upon by an external force. Every structure that can be seen to remain standing on a daily basis is in equilibrium; it is at rest and each of its members, combination of its members or any part of a member that is supporting a load are also at rest. There is a net result of zero in all directions for all of the applied loads and reactions.

CLARIFICATION

The intersection of the supply and demand curve establishes equilibrium. Equilibrium prices are prices that equate buyers' willingness to pay for the last units of output (prices along demand curve) with sellers' willingness to sell (cost represented by the supply curve) the last units of output. Thus, prices are said to represent marginal evaluations of goods and services. At this equilibrium price, there is neither a shortage or surplus in the market.

To better understand the concept of market equilibrium, suppose sellers produced Q1. To produce this level of output, sellers would have to receive a price equal to Ps. Sellers must receive Ps to cover the cost to produce the last units of output (remember the supply curve is an aggregate marginal cost curve). Buyers, however, are willing to pay only Pd for Q1. The sellers will accumulate a surplus of output unless prices (and output) are reduced. What if sellers produced Q2? To produce this level of output, sellers would only need to charge a price equal to Ps. Buyers, however, are willing to pay up to Pd for Q2. When consumers are willing to pay higher prices for additional output, but cannot buy the output, a shortage exists.

FREE BODY DIAGRAM

One of the most useful aids for solving a statics problem is the free body diagram (FBD). A free body diagram is a graphic, symbolic representation of the body (structure, element or piece of an element) with all connecting "parts" removed. The "parts" which have been removed are all of the real physical aspects of the structure. The body is represented by a simple line. Every "part" which has been removed is replaced by the external forces representing the internal forces present where that "part" connects with the other member in the FBD. Loads, distributed or concentrated, are removed and replaced with representational force systems. Everything that is needed to solve a force system is included on the FBD. Free body diagrams may not seem necessary in the relatively simple current applications, but as problems become more complex, their usefulness increases. The following is the process for determining the reaction at the wall for a cantilever beam. A FBD is first drawn of the beam. Next, cut the beam free from the wall and replace the wall with the forces that were supporting the beam at the wall before it was cut free. These forces are unknown, but they are the only forces that can keep the beam in equilibrium. They are identical to the internal forces in the beam at that point before it was cut.

A fixed support will resist translation in all directions and rotation (moment). The FBD must show all of these directions. The principles of equilibrium can always be used to solve a FBD. In the FBD above Sum Fy =2K and Sum Fx = 0. The 2K forces (load and vertical reaction force) cause a counter-clockwise couple of 10 K-FT which must be resisted by a moment on the end of the cut section of 10 K-FT acting in a clockwise direction. The following are three different systems which all have one 100 pound load and one 150 pound load actin on them at exactly the same point. Note how the internal arrangements of the three has been removed. The internal arrangement does not matter for the determination of the supporting reactions.

❐

5

Formulation of Quantum Statistical Mechanics

Quantum statistical mechanics is the study of statistical ensembles of quantum mechanical systems. A statistical ensemble is described by adensity operator S, which is a non-negative, self-adjoint, trace-class operator of trace 1 on the Hilbert space *H* describing the quantum system. This can be shown under various mathematical formalisms for quantum mechanics. One such formalism is provided by quantum logic.

EXPECTATION

From classical probability theory, we know that the expectation of a random variable X is completely determined by its distribution D_X by,

$$E(X) = \int_R \lambda dDx(\lambda)$$

assuming, of course, that the random variable is integrable or that the random variable is non-negative. Similarly, let A be an observable of a quantum mechanical system. A is given by a densely defined self-adjoint operator on H. The spectral measure of A defined by,

$$E_A(U) = \int_u \lambda dE(\lambda)$$

uniquely determines A and conversely, is uniquely determined by A. EA is a boolean homomorphism from the Borel subsets of R into the lattice Q of self- adjoint projections of H. In analogy with probability theory, given a state S, we introduce the distribution of A under S which is the probability measure defined on the Borel subsets of R by,

$$D_A\ (U) = Tr\ (E_A\ (U)S)$$

Similarly, the expected value of A is defined in terms of the probability distribution D_A by,

$$E(X) = \int_R \lambda dDx(\lambda)$$

Note that this expectation is relative to the mixed state S which is used in the definition of DA. Remark. For technical reasons, one needs to consider separately the positive and negative parts of A defined by the Borel functional calculus for unbounded operators.

One can easily show:

$$E(A) = Tr\ (AS) = Tr\ (SA)$$

Note that if S is a pure state corresponding to the vector ø,

$$E(A) = <\psi \mid A \mid \psi>$$

SYMMETRY REQUIREMENTS IN QUANTUM MECHANICS

Consider a gas consisting of *N identical*, non-interacting, structureless particles enclosed within a container of volume V. Let Qi denote collectively all the coordinates of the ith particle: i.e., the three Cartesian coordinates which determine its spatial position, as well as the spin coordinate which determines its internal state. Let si be an index labeling the possible quantum states of the ith particle: i.e., each possible value of Si corresponds to a specification of the three momentum components of the particle, as well as the direction of its spin orientation. According to quantum mechanics, the overall state of the system when the ith particle is in state si, etc., is completely determined by the *complex wave-function*,

$$\psi_{s1} \cdots, S_N (Q_1, Q_2, \cdots, Q_N)$$

In particular, the probability of an observation of the system finding the ith particle with coordinates in the range Qi to Qi + dQi, etc., is simply,

$$\psi_{s1} \cdots (Q_1, Q_2, \cdots, Q_N)^2\ dQ_1\ dQ_2\ \cdots Q_N$$

One of the fundamental postulates of quantum mechanics is the essential indistinguishability of particles of the same species. What this means, in practice, is that we cannot label particles of the same species: i.e., a proton is just a proton—we cannot meaningfully talk of proton number 1 and proton number 2, etc. Note that no such constraint arises in classical mechanics. Thus, in classical mechanics particles of the same species are regarded as being distinguishable, and can, therefore, be labelled. Of course, the quantum mechanical approach is the correct one.

Suppose that we *interchange* the ith and j[th] particles: i.e.,

$$Q_i = Q_j$$

$$S_i = S_j$$

If the particles are truly indistinguishable then nothing has changed: i.e., we have a particle in quantum state si and a particle in quantum state si both before and after the particles are swapped. Thus, the probability of observing the system in a given state also cannot have changed: i.e.,

$$\left|\psi\left(\cdots Qi\cdots Qj\cdots\right)\right|^2 = \left|\psi\left(\cdots Q_j\cdots Q_i\right)\right|^2$$

Here, we have omitted the subscripts s1..., sN for the sake of clarity. Note that we cannot conclude that the wave-function ψ is unaffected when the particles are swapped, because ψ cannot be observed experimentally. Only the probability density ψ is observable. Equation implies that,

$$\left|\psi\left(\cdots Qi\cdots Qj\cdots\right)\right| = \left|A\psi\left(\cdots Q_j\cdots Q_i\right)\right|$$

where A is a complex constant of modulus unity: i.e. A1ψ .

Suppose that we interchange the ith and jth particles a second time. Swapping the ith and jth particles twice leaves the system completely unchanged: i.e., it is equivalent to doing nothing to the system. Thus, the wave- functions before and after this process must be identical. It follows from Equation that,

$$A^2 = 1.$$

Of course, the only solutions to the above equation are A= ±1.

We conclude, from the above discussion, that the wave-function □ is either completely *symmetric* under the interchange of particles, or it is completely *anti-symmetric*. In other words, either,

$$\psi(\ldots Q_i \ldots Q_j \ldots) = +\,\psi\,(\ldots Q_j \ldots Q_i \ldots).$$

or,

$$\psi(\ldots Qi \ldots Qj \ldots) = -\psi\,(\ldots Qj \ldots Qi \ldots).$$

In 1940 the Nobel prize winning physicist Wolfgang Pauli demonstrated, via arguments involving relativistic invariance, that the wave-function associated with a collection of identical integer-spin (i.e., spin 0, 1, 2, etc.) particles satisfies, whereas the wave-function associated with a collection of identical half-integer-spin (i.e., spin 1/2, 3/2, 5/2, etc.) particles satisfies. The former type of particles are known as bosons. The latter type of particles are called fermions (after the Italian physicists Enrico Fermi, who first studied the properties of fermion gases). Common examples of bosons are photons and He^4 atoms. Common examples of fermions are protons, neutrons, and electrons.

Consider a gas made up of identical bosons. Equation implies that the interchange of any two particles does not lead to a new state of the system. Bosons must, therefore, be considered as genuinely indistinguishable when enumerating the different possible states of the gas. Note that Equation above imposes no restriction on how many particles can occupy a given single-particle quantum state s.

Consider a gas made up of identical fermions. Equation implies that the interchange of any two particles does not lead to a new physical state of the system (since ψ is invariant). Hence, fermions must also be considered genuinely indistinguishable when enumerating the different possible states of the gas. Consider the special case where particles i and j lie in the same quantum state. In this case, the act of swapping the two particles is equivalent to leaving the system unchanged, so,

$$\psi(\ldots Q_i \ldots Q_j \ldots) = \psi(\ldots Q_j \ldots Q_i \ldots).$$

However, Equation above is also applicable, since the two particles are fermions. The only way in which Equations above can be reconciled is if,

$$\psi = 0$$

wherever particles i and j lie in the same quantum state. This is another way of saying that it is impossible for any two particles in a gas of fermions to lie in the same single-particle quantum state. This proposition is known as the Pauli exclusion principle, since it was first proposed by W. Pauli in 1924 on empirical grounds.

Consider, for the sake of comparison, a gas made up of identical classical particles. In this case, the particles must be considered distinguishable when enumerating the different possible states of the gas. Furthermore, there are no constraints on how many particles can occupy a given quantum state.

There are three different sets of rules which can be used to enumerate the states of a gas made up of identical particles. For a boson gas, the particles must be treated as being indistinguishable, and there is no limit to how many particles can occupy a given quantum state. This set of rules is called Bose- Einstein statistics, after S.N.

Bose and A. Einstein, who first developed them. For a fermion gas, the particles must be treated as being indistinguishable, and there can never be more than one particle in any given quantum state. This set of rules is called Fermi-Dirac statistics, after E. Fermi and P.A.M. Dirac, who first developed them. Finally, for a classical gas, the particles must be treated as being distinguishable, and there is no limit to how many particles can occupy a given quantum state. This set of rules is called *Maxwell-Boltzmann statistics,* after J.C. Maxwell and L. Boltzmann, who first developed them.

An Illustrative Example

Consider a very simple gas made up of two identical particles. Suppose that each particle can be in one of three possible quantum states, s = 1, 2, 3. Let us enumerate the possible states of the whole gas according to Maxwell-Boltzmann, Bose-Einstein, and Fermi-Dirac statistics, respectively.

For the case of Maxwell-Boltzmann (MB) statistics, the two particles are considered to be distinguishable. Let us denote them A and B. Furthermore, any number of particles can occupy the same quantum state.

Table. Two Particles Distributed Amongst Three States According to Maxwell-Boltzmann statistics.

1	2	3
AB	...	...
...	AB	...
...	...	AB
A	B	...
B	A	...
A	...	B
B	...	A
...	A	B
...	B	A

There are clearly 9 distinct states.

For the case of Bose-Einstein (BE) statistics, the two particles are considered to be indistinguishable.

Let us denote them both as A Furthermore, any number of particles can occupy the same quantum state. The possible different states of the gas are shown in Table.

Table. Two Particles Distributed Amongst Three States According to Bose-Einstein Statistics.

1	2	3
AA	...	...
...	AA	...
...	...	AA
A	A	...
A	...	A
...	A	A

There are clearly 6 distinct states.

Finally, for the case of Fermi-Dirac (FD) statistics, the two particles are considered to be indistinguishable. Let us again denote them both as A. Furthermore, no more than one particle can occupy a given quantum state. The possible different states of the gas are shown in Table.

Table. Two Particles Distributed Amongst Three States According to Fermi-Dirac Statistics.

1	2	3
A	A	...
A	...	A
...	A	A

There are clearly only 3 distinct states. It follows, from the above example, that Fermi-Dirac statistics are more restrictive (i.e., there are less possible states of the system) than Bose-Einstein statistics, which are, in turn, more restrictive than Maxwell-Boltzmann statistics. Let, Probabitity that the two particles are found in the same state.

Probability that the two particles are found in different states

For the case under investigation,

$$\xi MB = 1 / 2$$

$$\xi BE = 1,$$

$$\xi FD = 0.$$

We conclude that in Bose-Einstein statistics there is a greater relative tendency for particles to cluster in the same state than in classical statistics. On the other hand, in Fermi-Dirac statistics there is less tendency for particles to cluster in the same state than in classical statistics.

FORMULATION OF THE STATISTICAL PROBLEM

Consider a gas consisting of N identical non-interacting particles occupying volume V and in thermal equilibrium at temperature T. Let us label the possible quantum states of a single particle by r (or s). Let the energy of a particle in state r be denoted ξr. Let the number of particles in state r be written nr. Finally, let us label the possible quantum states of the whole gas by R.

The particles are assumed to be non-interacting, so the total energy of the gas in state R, where there are nr particles in quantum state r, etc., is simply,

$$E_R = \xi n,$$

where the sum extends over all possible quantum states T. Furthermore, since the total number of particles in the gas is known to be N.

In order to calculate the thermodynamic properties of the gas (i.e., its internal energy or its entropy), it is necessary to calculate its partition function,

$$Z = \sum_R e^{-\beta E} R = \sum_R e^{-\beta}\left(n_1 c_1 + c_2 + \cdots\right)$$

Here, the sum is over all possible states R of the whole gas: i.e., over all the various possible values of the numbers n_1, n_2,.... .

FERMI-DIRAC STATISTICS

Let us, first of all, consider Fermi-Dirac statistics. According to Equation, the average number of particles in quantum state s can be written,

$$\overline{n}_s = \frac{\Sigma n_s n_s e^{-\beta nscs} \sum_{n1,n2}^{(s)} e^{-\beta}\left(n_1 c_1 + n_2 c_2 + \cdots\right)}{\Sigma_{ns} e^{-\beta nscs} \sum_{n1,n2}^{(s)} e^{-\beta}\left(n_1 + c_1 + n_2 c_2 + \cdots\right)}$$

Here, we have rearranged the order of summation, using the multiplicative properties of the exponential function. Note that the first sums in the numerator and denominator only involve ns, whereas the last sums omit the particular state s from consideration (this is indicated by the superscript s on the summation symbol). Of course, the sums in the above expression range over all values of the numbers $n_1 n_2 \ldots$ such that $n_r = 0$ and 1 for each r, subject to the overall constraint that,

$$\sum_r n_r = N$$

Let us introduce the function,

$$Zs(N) = \sum_{n_1 n_2}^{(s)} e^{-\beta}(n_1 e_1 + n_2 c_2 + \cdots)$$

which is defined as the partition function for N particles distributed over all quantum states, excluding state s, according to Fermi-Dirac statistics. By explicitly performing the sum over ns = 0 and 1, the expression reduces to,

$$Z = \sum_R e^{-\beta E} R = \sum_R e^{-\beta}(n_1 c_1 + c_2 + \cdots)$$

which yields,

$$\bar{n}_s = \frac{\Sigma n_s n_s e^{-\beta nscs} \sum_{n1,n2}^{(s)} e^{-\beta}(n_1 c_1 + n_2 c_2 + \cdots)}{\Sigma_{ns} e^{-\beta nscs} \sum_{n1,n2}^{(s)} e^{-\beta}(n_1 + c_1 + n_2 c_2 + \cdots)}$$

In order to make further progress, we must somehow relate Z ξ Nξ 1ξ can be Taylor

$$\sum_r n_r = N$$

As always, we Taylor expand the slowly varying function In Zs (N), rather than the rapidly varying function Zs (N), because the radius of convergence of the latter Taylor series is too small for the series to be of any practical use.

Now, since Zs (N) is a sum over very many different quantum states, we would not expect the logarithm of this function to be sensitive to which particular state s is excluded from consideration. Let us, therefore, introduce the approximation that ξs is independent of s, so that we can write, for all s. It follows that the derivative can be expressed approximately in terms of the derivative of the full partition function Z(N) (in which the N particles are distributed over all quantum states). In fact,

$$\left|\psi\left(\cdots Qi\cdots Qj\cdots\right)\right|^2=\left|\psi\left(\cdots Q_j\cdots Q_i\right)\right|^2$$

Making use of Equation, with ΔN=1, plus the approximation, the expression reduces to,

$$\left|\psi\left(\cdots Qi\cdots Qj\cdots\right)\right|=\left|A\psi\left(\cdots Q_j\cdots Q_i\right)\right|$$

This is called the Fermi-Dirac distribution. The parameter ξ is determined by the constraint that ξ nr ξ N : i.e.,

$$\psi_{s1}\cdots,S_N\left(Q_1,Q_2,\cdots,Q_N\right)$$

becomes sufficiently large. On the other hand, since the denominator in Equation can never become less than unity, no matter how small ξs becomes, it follows that ns1.

Thus, $0<n_s$, $n_s<1$, in accordance with the Pauli exclusion principle. Equations above can be integrated to give,

$$In\ Z=\alpha N+\sum_r In\left(1+e^{-\alpha-ber}\right)$$

Photon Statistics

Up to now, we have assumed that the number of particles N contained in a given system is a fixed number. This is a reasonable assumption if the particles possess non-zero mass, since we are not generally considering relativistic systems in this course. However, this assumption breaks down for the case of photons, which are zero-mass bosons. In fact, photons enclosed in a container of volume V, maintained at temperature T, can readily be absorbed or emitted by the walls. Thus, for the special case of a gas of photons there is no requirement which limits the total number of particles. It follows, from the above discussion, that photons obey a simplified form of Bose-Einstein statistics in which there is an unspecified total number of particles. This type of statistics is called photon statistics. Consider the expression. For the case of photons, the numbers n1, n2... assume all values nr = 0, 1, 2... for each r, without any further restriction.

$$\overline{n}_s=\frac{\Sigma_{ns}n_s e^{-\beta N_s e_s}}{\Sigma_{ns}e^{-\beta n_s e_s}}$$

It follows that the sums in the numerator and denominator are identical and, therefore, cancel. Hence, Equation reduces to,

$$\overline{n}_s = -\frac{1}{\beta}\frac{\partial}{\partial \in_s}\left(In \sum_{N_s} e^{-\beta_{NS} e_s} \right)$$

However, the above expression can be rewritten,

$$\sum_{N_s=0}^{\infty} e^{-\beta N_s e_s} = 1 + e^{-\beta e_s} + \cdots = \frac{1}{1-e^{-\beta e_s}}$$

Now, the sum on the right-hand side of the above equation is an infinite geometric series, which can easily be evaluated. In fact,

$$\overline{n}_s \frac{1}{\beta}\frac{\partial}{\partial \in_s} In\left(1-e^{-\beta e_s}\right) = \frac{e^{-\beta s}}{1-e^{-\beta e_s}}$$

This is known as the Planck distribution, after the German physicist Max Planck who first proposed it in 1900 on purely empirical grounds.

Equation can be integrated to give,

$$In\, Z = \alpha N + \sum_r In\left(1+e^{-\alpha-ber}\right)$$

where use has been made of Equation.

BOSE-EINSTEIN STATISTICS

Let us now consider Bose-Einstein statistics. The particles in the system are assumed to be massive, so the total number of particles N is a fixed number. Consider the expression. For the case of massive bosons, the numbers n_1, n_2...assume all values n_r = 0, 1, 2... for each r, subject to the constraint that r nr = N . Performing explicitly the sum over ns, this expression reduces to,

$$Z = \sum_R e^{-\beta E} R = \sum_R e^{-\beta}\left(n_1 c_1 + c_2 + \cdots\right)$$

where Zs (N) is the partition function for N particles distributed over all quantum states, excluding state s, according to Bose-Einstein statistics. Using Equation, and the approximation, the above equation reduces to,

$$\overline{n}_s = \frac{\Sigma n_s n_s e^{-\beta nscs} \sum_{n1,n2}^{(s)} e^{-\beta}\left(n_1 c_1 + n_2 c_2 + \cdots\right)}{\Sigma_{ns} e^{-\beta nscs} \sum_{n1,n2}^{(s)} e^{-\beta}\left(n_1 + c_1 + n_2 c_2 + \cdots\right)}$$

Note that this expression is identical to equation, except that is replaced by s. Hence, an analogous calculation to that outlined in the previous subsection yields,

$$\sum_r n_r = N$$

This is called the Bose-Einstein distribution. Note that ns large in this distribution.

$$Zs(N) = \sum_{n_1 n_2}^{(s)} e^{-\beta}(n_1 e_1 + n_2 c_2 + \cdots)$$

The parameter ψ is again determined by the constraint on the total number of particles: i.e.,

$$\bar{n}_s = \frac{\Sigma_{ns} n_s e^{-\beta_{N_s} e_s}}{\Sigma_{ns} e^{-\beta n_s e_s}}$$

Equations above can be integrated to give,

$$In\, Z = \alpha N + \sum_r In\left(1 + e^{-\alpha - ber}\right)$$

where use has been made of Equation.

Note that photon statistics correspond to the special case of Bose-Einstein statistics in which the parameter a takes the value zero, and the constraint does not apply.

MAXWELL-BOLTZMANN STATISTICS

For the purpose of comparison, it is instructive to consider the purely classical case of Maxwell-Boltzmann statistics. The partition function is written,

$$Z = \sum_R e^{-\beta(n_1 e_1 + n_2 e_2 + \cdots)}$$

where the sum is over all distinct states R of the gas, and the particles are treated as distinguishable. For given values of n1,n2 there are,

$$\frac{N!}{n_1!n_2!\ldots}$$

possible ways in which N distinguishable particles can be put into individual quantum states such that there are n1 particles in state 1, n2 particles in state 2, etc. Each of these possible arrangements corresponds to a distinct state for the whole gas. Hence, Equation can be written,

$$Z = \sum_{n_1 n_2} \frac{N!}{n_1!n_2!...} e^{-\beta(n_1 e_1 + n_2 e_2 + ...)}$$

where the sum is over all values of nr = 0,1,2... for each r, subject to the constraint that,

$$\sum_r n_r = N$$

Now, Equation (above) can be written,

$$Z = \sum_R e^{-\beta(n_1 e_1 + n_2 e_2 + \cdots)}$$

which, by virtue of Equation, is just the result of expanding a polynomial. In fact,

$$\frac{N!}{n_1!n_2!...}$$

or,

$$Z = \sum_{n_1 n_2} \frac{N!}{n_1!n_2!} \left(e^{-\beta e1}\right) n1 \left(e^{-\beta e_2}\right) n_2 ...$$

$$Z = \left(e^{-\beta e1} + e^{-\beta e1} + e^{-be2} + ...\right) N$$

Note that the argument of the logarithm is simply the partition function for a single particle.

Equations above can be combined to give,

$$In\, Z = N\ In \left(\sum_r -\beta^{er} \right)$$

This is known as the Maxwell-Boltzmann distribution. It is, of course, just the result obtained by applying the Boltzmann distribution to a single particle.

Quantum Statistics in the Classical Limit

The preceding analysis regarding the quantum statistics of ideal gases is summarized in the following statements. The mean number of particles occupying quantum state s is given by,

$$Z = \sum_{R} e^{-\beta(n_1 e_1 + n_2 e_2 + \cdots)}$$

where the upper sign corresponds to Fermi-Dirac statistics and the lower sign corresponds to Bose-Einstein statistics. The parameter a is determined via,

$$\frac{N!}{n_1!n_2!\ldots}$$

Finally, the partition function of the gas is given by,

$$Z = \sum_{n_1 n_2} \frac{N!}{n_1!n_2!\ldots} e^{-\beta(n_1 e_1 + n_2 e_2 + \ldots)}$$

Let us investigate the magnitude of in some important limiting cases. Consider, first of all, the case of a gas at a given temperature when its concentration is made sufficiently low: i.e., when N is made sufficiently small. The relation can only be satisfied if each term in the sum over states is made sufficiently small; i.e., if nr exp (a+b) or exp exp (a+b) for all states r. Consider, next, the case of a gas made up of a fixed number of particles when its temperature is made sufficiently large: i.e., when r is made sufficiently small. In the sum in Equation, the terms of appreciable magnitude are those for which exp (a+b). Thus, it follows that as an increasing number of terms with large values of r contribute substantially to this sum. In order to prevent the sum from exceeding N, the parameter must become large enough that each term is made sufficiently small: i.e., it is again necessary that nr1 or exp (a+b)for all states r.

The above discussion suggests that if the concentration of an ideal gas is made sufficiently low, or the temperature is made sufficiently high, then must become so large that,

$$e^{a+brr}>>1$$

for all r. Equivalently, this means that the number of particles occupying each quantum state must become so small that,

$$e^{a+brr}>>1$$

for all r. It is conventional to refer to the limit of sufficiently low concentration, or sufficiently high temperature, in which Equations above are satisfied, as the classical limit.

According to Equations above, both the Fermi-Dirac and Bose-Einstein distributions reduce to,

$$In\,Z = \alpha N + \sum_r In\left(1 + e^{-\alpha - ber}\right)$$

in the classical limit, whereas the constraint yields,

$$\overline{n}_s = \frac{\Sigma_{ns} n_s e^{-\beta_{N_s} e_s}}{\Sigma_{ns} e^{-\beta n_s e_s}}$$

The above expressions can be combined to give,

$$\overline{n}_s = -\frac{1}{\beta}\frac{\partial}{\partial \in_s}\left(In \sum_{N_s} e^{-\beta_{NS} e_s}\right)$$

It follows that in the classical limit of sufficiently low density, or sufficiently high temperature, the quantum distribution functions, whether Fermi-Dirac or Bose-Einstein, reduce to the Maxwell-Boltzmann distribution. It is easily demonstrated that the physical criterion for the validity of the classical approximation is that the mean separation between particles should be much greater than their mean de Broglie wavelengths. Let us now consider the behaviour of the partition function in the classical limit. We can expand the logarithm to give,

$$\frac{N!}{n_1!n_2!\ldots}$$

However, according to Equation above,

$$Z = \sum_{n_1 n_2} \frac{N!}{n_1!n_2!\ldots} e^{-\beta(n_1 e_1 + n_2 e_2 + \ldots)}$$

It follows that,

$$\sum_r n_r = N$$

Note that this does not equal the partition function ZMB computed in Equation from Maxwell-Boltzmann statistics: i.e.,

$$\overline{n}_s = N\frac{e - \beta e_s}{\Sigma_r e - \beta e_r}$$

where use has been made of Stirling's approximation (N ! □ N In N – N), since N is large. Here, the factor N! simply corresponds to the number of different permutations of the N particles: permutations which are physically meaningless when the particles are identical. Recall, that we had to introduce precisely this factor, in an ad hoc fashion order to avoid the non-physical consequences of the Gibb's paradox. Clearly, there is no Gibb's paradox when an ideal gas is treated properly via quantum mechanics.

In the classical limit, a full quantum mechanical analysis of an ideal gas reproduces the results obtained, except that the arbitrary parameter h0 is replaced by Planck's constant h = 6.61 × 10–34 Js. A gas in the classical limit, where the typical de Broglie wavelength of the constituent particles is much smaller than the typical inter-particle spacing, is said to be non-degenerate. In the opposite limit, where the concentration and temperature are such that the typical de Broglie wavelength becomes comparable with the typical inter-particle spacing, and the actual Fermi-Dirac or Bose-Einstein distributions must be employed, the gas is said to be degenerate.

SPECIFICATION OF THE STATE OF A MANY PARTICLE SYSTEM

How do we determine the state of a many particle system? Well, let us, first of all, consider the simplest possible many particle system, which consists of a single spinless particle moving classically in one dimension. Assuming that we know the particle's equation of motion, the state of the system is fully specified once we simultaneously measure the particle's position q and momentum p. In principle, if we know q and p then we can calculate the state of the system at all subsequent times using the equation of motion. In practice, it is impossible to specify q and p exactly, since there is always an intrinsic error in any experimental measurement.

Consider the time evolution of q and p. This can be visualised by plotting the point (q, p) in the q-p plane. This plane is generally known as phase-space. In general, the point (q, p) will trace out some very complicated pattern in phase-space. Suppose that we divide phase-space into rectangular cells of uniform dimensions q and p. Here, q is the intrinsic error in the position measurement, and q the intrinsic error in the momentum measurement. The "area" of each cell is,

$$\Sigma q \, \Sigma p = h_0$$

where h_0 is a small constant having the dimensions of angular momentum. The coordinates q and p can now be conveniently specified by indicating the cell in phase-space into which they plot at any given time. This procedure automatically ensures that we do not attempt to specify q and p to an accuracy greater than our experimental error, which would clearly be pointless. Let us now consider a single spinless particle moving in three dimensions. In order to specify the state of the system we now need to know three q-p pairs: i.e., qxΣ – px, qy – py, and qz – pz. Incidentally, the number of q-p pairs needed to specify the state of the system is usually called the number of degrees of freedom of the system. Thus, a single particle moving in one dimension constitutes a one degree of freedom system, whereas a single particle moving in three dimensions constitutes a three degree of freedom system.

Consider the time evolution of q and p, where q = (qx, qy, qz), etc. This can be visualised by plotting the point (q, p) in the six dimensional q – p phase- space. Suppose that we divide the qx-px plane into rectangular cells of uniform dimensions Σq and Σp, and do likewise for the qy – py and qy – pz planes. Here, Σq and Σp are again the intrinsic errors in our measurements of position and momentum, respectively. This is equivalent to dividing phase-space up into regular six dimensional cells of volume h03. The coordinates q and p can now be conveniently specified by indicating the cell in phase-space into which they plot at any given time. Again, this procedure automatically ensures that we do not attempt to specify q and p to an accuracy greater than our experimental error.

Finally, let us consider a system consisting of N spinless particles moving classically in three dimensions. In order to specify the state of the system, we need to specify a large number of q – p pairs. The requisite number is simply the number of degrees of freedom, f. For the present case, f = 3N. Thus, phase- space (i.e., the space of all the q-p pairs) now possesses 2f = 6 N dimensions. Consider a particular pair of conjugate coordinates, qi and pi. As before, we divide the Σq-Σp plane into rectangular cells of uniform dimensions Σq and Σp.

This is equivalent to dividing phase-space into regular 2f dimensional cells of volume h0f. The state of the system is specified by indicating which cell it occupies in phase-space at any given time.

In principle, we can specify the state of the system to arbitrary accuracy by taking the limit hoΣ0. In reality, we know from quantum mechanics that it is impossible to simultaneously measure a coordinate qi and its conjugate momentum pi to greater accuracy than ΣqiΣpi = Σ . This implies that, h_o > qΣ.

In other words, the uncertainty principle sets a lower limit on how finely we can chop up classical phase-space.

In quantum mechanics we can specify the state of the system by giving its wave-function at time t,

$$\Sigma q1 ,..., q f , s1 ...sg , t ,$$

where f is the number of translational degrees of freedom, and g the number of internal (e.g., spin) degrees of freedom. For instance, if the system consists of N spin-one-half particles then there will be 3N translational degrees of freedom, and N spin degrees of freedom (i.e., the spin of each particle can either point up or down along the z-axis). Alternatively, if the system is in a stationary state (i.e., an eigenstate of the Hamiltonian) then we can just specify f+g quantum numbers. Either way, the future time evolution of the wave- function is fully determined by Schrödinger's equation.

In reality, this approach does not work because the Hamiltonian of the system is only known approximately. Typically, we are dealing with a system consisting of many weakly interacting particles. We usually know the Hamiltonian for completely non-interacting particles, but the component of the Hamiltonian associated with particle interactions is either impossibly complicated, or not very well known (often, it is both!). We can define approximate stationary eigenstates using the Hamiltonian for non-interacting particles. The state of the system is then specified by the quantum numbers identifying these eigenstates. In the absence of particle interactions, if the system starts off in a stationary state then it stays in that state for ever, so its quantum numbers never change. The interactions allow the system to make transitions between different "stationary" states, causing its quantum numbers to change in time.

THE PRINCIPLE OF EQUAL A PRIORI PROBABILITIES

We now know how to specify the instantaneous state of a many particle system. In principle, such a system is completely deterministic. Once we know the initial state and the equations of motion (or the Hamiltonian) we can evolve the system forward in time and, thereby, determine all future states. In reality, it is quite impossible to specify the initial state or the equations of motion to sufficient accuracy for this method to have any chance of working. Furthermore, even if it were possible, it would still not be a practical proposition to evolve the equations of motion.

Remember that we are typically dealing with systems containing Avogadro's number of particles: i.e., about 1024 particles. We cannot evolve 1024 simultaneous differential equations! Even if we could, we would not want to. After all, we are not particularly interested in the motions of individual particles. What we really want is statistical information regarding the motions of all particles in the system.

Clearly, what is required here is a statistical treatment of the problem. Instead of focusing on a single system, let us proceed in the usual manner and consider a statistical ensemble consisting of a large number of identical systems. In general, these systems are distributed over many different states at any given time. In order to evaluate the probability that the system possesses a particular property, we merely need to find the number of systems in the ensemble which exhibit this property, and then divide by the total number of systems, in the limit as the latter number tends to infinity. We can frequently place some general constraints on the system. Typically, we know the total energy E, the total volume V, and the total number of particles N. To be more honest, we can only really say that the total energy lies between E and E + SE, etc., where SE is an experimental error. Thus, we only need concern ourselves with those systems in the ensemble exhibiting states which are consistent with the known constraints. We call these the states accessible to the system. In general, there are a great many such states. We now need to calculate the probability of the system being found in each of its accessible states. Well, perhaps "calculate" is the wrong word. The only way we could calculate these probabilities would be to evolve all of the systems in the ensemble and observe how long on average they spend in each accessible state. But, as we have already mentioned, such a calculation is completely out of the question. So what do we do instead? Well, we effectively guess the probabilities.

Permit us consider an isolated system in equilibrium. In this situation, we would expect the probability of the system being found in one of its accessible states to be independent of time. This implies that the statistical ensemble does not evolve with time. Individual systems in the ensemble will constantly change state, but the average number of systems in any given state should remain constant. Thus, all macroscopic parameters describing the system, such as the energy and the volume, should also remain constant. There is nothing in the laws of mechanics which would lead us to suppose that the system will be found more often in one of its accessible states than in another. We assume, therefore, that the system is equally likely to be found in any of its accessible states. This is called the assumption of equal a priori probabilities, and lies at the very heart of statistical mechanics.

In fact, we use assumptions like this all of the time without really thinking about them. Suppose that we were asked to pick a card at random from a well-shuffled pack. I think that most people would accept that we have an equal probability of picking any card in the pack. There is nothing which would favour one particular card over all of the others. So, since there are fifty-two cards in a normal pack, we would expect the probability of picking the Ace of Spades, say, to be 1/52.

We could now place some constraints on the system. For instance, we could only count red cards, in which case the probability of picking the Ace of Hearts, say, would be 1/26, by the same reasoning. In both cases, we have used the principle of equal a priori probabilities. People really believe that this principle applies to games of chance such as cards, dice, and roulette. In fact, if the principle were found not to apply to a particular game most people would assume that the game was "crooked." But, imagine trying to prove that the principle actually does apply to a game of cards. This would be very difficult! We would have to show that the way most people shuffle cards is effective at randomising their order. A convincing study would have to be part mathematics and part psychology!

In statistical mechanics, we treat a many particle system a bit like an extremely large game of cards. Each accessible state corresponds to one of the cards in the pack. The interactions between particles cause the system to continually change state. This is equivalent to constantly shuffling the pack. Finally, an observation of the state of the system is like picking a card at random from the pack. The principle of equal a priori probabilities then boils down to saying that we have an equal chance of choosing any particular card.

It is, unfortunately, impossible to prove with mathematical rigour that the principle of equal a priori probabilities applies to many-particle systems. Over the years, many people have attempted this proof, and all have failed miserably. Not surprisingly, therefore, statistical mechanics was greeted with a great deal of scepticism when it was first proposed just over one hundred years ago. One of the its main proponents, Ludvig Boltzmann, got so fed up with all of the criticism that he eventually threw himself off a bridge! Nowadays, statistical mechanics is completely accepted into the cannon of physics. The reason for this is quite simple: it works! It is actually possible to formulate a reasonably convincing scientific case for the principle of equal a priori probabilities. To achieve this we have to make use of the so-called H theorem.

The H Theorem

Consider a system of weakly interacting particles. In quantum mechanics we can write the Hamiltonian for such a system as,

$$H = H_0 + H_1,$$

where H_0 is the Hamiltonian for completely non-interacting particles, and H_1 is a small correction due to the particle interactions. We can define approximate stationary eigenstates of the system using H_0. Thus,

$$H_0 \Sigma r \;\; \Sigma Er \, \Sigma r \, ,$$

where the index r labels a state of energy Er and eigenstate Σr. In general, there are many different eigenstates with the same energy: these are called degenerate states. Used for example, consider N non-interacting spinless particles of mass m confined in a cubic box of dimension L.

According to standard wave-mechanics, the energy levels of the ith particle are given by,

$$\overline{n}_s = N\frac{e - \beta e_s}{\Sigma_r e - \beta e_r}$$

where n_i1, n_i2, and n_i3 are three (positive integer) quantum numbers. The overall energy of the system is the sum of the energies of the individual particles, so that for a general state r,

The overall state of the system is thus specified by 3N quantum numbers (i.e., three quantum numbers per particle). There are clearly very many different arrangements of these quantum numbers which give the same overall energy.

Consider, now, a statistical ensemble of systems made up of weakly interacting particles. Suppose that this ensemble is initially very far from equilibrium. For instance, the systems in the ensemble might only be distributed over a very small subset of their accessible states. If each system starts off in a particular stationary state (i.e., with a particular set of quantum numbers) then, in the absence of particle interactions, it will remain in that state for ever. Hence, the ensemble will always stay far from equilibrium, and the principle of equal a priori probabilities will never be applicable. In reality, particle interactions cause each system in the ensemble to make transitions between its accessible "stationary" states. This allows the overall state of the ensemble to change in time.

Let us label the accessible states of our system by the index r. We can ascribe a time dependent probability Pr (t) of finding the system in a particular approximate stationary state r at time t. Of course, Pr (t) is proportional to the number of systems in the ensemble in state r at time t. In general, Pr is time dependent because the ensemble is evolving towards an equilibrium state.

We assume that the probabilities are properly normalised, so that the sum over all accessible states always yields,

Small interactions between particles cause transitions between the approximate stationary states of the system. There then exists some transition probability per unit time Wrs that a system originally in state r ends up in state s as a result of these interactions. Likewise, there exists a probability per unit time Wrs that a system in state r makes a transition to state s. These transition probabilities are meaningful in quantum mechanics provided that the particle interaction strength is sufficiently small, there is a nearly continuous distribution of accessible energy levels, and we consider time intervals which are not too small. These conditions are easily satisfied for the types of systems usually analysed via statistical mechanics (e.g., nearly ideal gases). One important conclusion of quantum mechanics is that the forward and inverse transition probabilities between two states are the same, so that,

$$Wrs = Wst$$

for any two states r and s. This result follows from the time reversal symmetry of quantum mechanics. On the microscopic scale of individual particles, all fundamental laws of physics (in particular, classical and quantum mechanics) possess this symmetry. So, if a certain motion of particles satisfies the classical equations of motion (or Schrödinger's equation) then the reversed motion, with all particles starting off from their final positions and then retracing their paths exactly until they reach their initial positions, satisfies these equations just as well.

Suppose that we were to "film" a microscopic process, such as two classical particles approaching one another, colliding, and moving apart. We could then gather an audience together and show them the film. To make things slightly more interesting we could play it either forwards or backwards. Because of the time reversal symmetry of classical mechanics, the audience would not be able to tell which way the film was running (unless we told them!). In both cases, the film would show completely plausible physical events.

We can play the same game for a quantum process. For instance, we could "film" a group of photons impinging on some atoms. Occasionally, one of the atoms will absorb a photon and make a transition to an "excited" state (i.e., a state with higher than normal energy). We could easily estimate the rate constant for this process by watching the film carefully. If we play the film backwards then it will appear to show excited atoms occasionally emitting a photon and decaying back to their unexcited state. If quantum mechanics possesses time reversal symmetry (which it certainly does!) then both films should appear equally plausible. This means that the rate constant for the absorption of a photon to produce an excited state must be the same as the rate constant for the excited state to decay by the emission of a photon. Otherwise, in the backwards film the excited atoms would appear to emit photons at the wrong rate, and we could then tell that the film was being played backwards. It follows, therefore, that as a consequence of time reversal symmetry, the rate constant for any process in quantum mechanics must equal the rate constant for the inverse process.

The probability Pr of finding the systems in the ensemble in a particular state r changes with time for two reasons. Firstly, systems in another state s can make transitions to the state r.

The rate at which this occurs is Ps, the probability that the systems are in the state s to begin with, times the rate constant of the transition Wst. Secondly, systems in the state r can make transitions to other states such as s. The rate at which this occurs is clearly Pr times Wrs. We can write a simple differential equation for the time evolution of Pr:

Where use has been made of the symmetry condition. The summation is over all accessible states.

Consider now the quantity H (from which the H theorem derives its name), which is the mean value of In Pr over all accessible states:

$$In\, Z = N\, In\left(\sum_r -\beta^{er}\right)$$

This quantity changes as the individual probabilities Pr vary in time. Straightforward differentiation of the above equation yields,

We can now interchange the dummy summations indices r and s to give,

$$Z = \sum_R e^{-\beta E} R = \sum_R e^{-\beta}\left(n_1 c_1 + c_2 + \cdots\right)$$

We can write dH/dt in a more symmetric form by adding the previous two equations and making use of Equation:

$$\overline{n}_s = \frac{\Sigma n_s n_s e^{-\beta nscs} \sum_{n1,n2}^{(s)} e^{-\beta}\left(n_1 c_1 + n_2 c_2 + \cdots\right)}{\Sigma_{ns} e^{-\beta nscs} \sum_{n1,n2}^{(s)} e^{-\beta}\left(n_1 + c_1 + n_2 c_2 + \cdots\right)}$$

Note, however, that In Pr is a monotonically increasing function of Pr. It follows that In Pr > In Ps whenever Pr > Ps, and vice versa. Thus, in general, the right-hand side of the above equation is the sum of many negative contributions.

Hence, we conclude that,

$$\sum_r n_r = N$$

The equality sign only holds in the special case where all accessible states are equally probable, so that Pr = Ps for all r and s. This result is called the H theorem, and was first proved by the unfortunate Professor Boltzmann.

The H theorem tells us that if an isolated system is initially not in equilibrium then it will evolve under the influence of particle interactions in such a manner that the quantity H always decreases. This process will continue until H reaches its minimum possible value, at which point dH/dt = 0, and there is no further evolution of the system. According to Equation above, in this final equilibrium state the system is equally likely to be found in any one of its accessible states. This is, of course, the situation predicted by the principle of equal a priori probabilities.

You may be wondering why the above argument does not constitute a mathematically rigourous proof that the principle of equal a priori probabilities applies to many particle systems. The answer is that we tacitly made an unwarranted assumption: i.e., we assumed that the probability of the system making a transition from some state r to another state s is independent of the past history of the system.

In general, this is not the case in physical systems, although there are many situations in which it is a pretty good approximation. Thus, the epistemological status of the principle of equal a priori probabilities is that it is plausible, but remains unproven. As we have already mentioned, the ultimate justification for this principle is empirical: i.e., it leads to theoretical predictions which are in accordance with experimental observations.

The Relaxation Moment

The H theorem guarantees that an isolated many particle system will eventually reach equilibrium, irrespective of its initial state. The typical time- scale for this process is called the relaxation time, and depends in detail on the nature of the inter-particle interactions. The principle of equal a priori probabilities is only valid for equilibrium states. It follows that we can only safely apply this principle to systems which have remained undisturbed for many relaxation times since they were setup, or last interacted with the outside world. The relaxation time for the air in a typical classroom is very much less than one second.

This suggests that such air is probably in equilibrium most of the time, and should, therefore, be governed by the principle of equal a priori probabilities. In fact, this is known to be the case. Consider another example. Our galaxy, the "Milky Way," is an isolated dynamical system made up of about 1011 stars. In fact, it can be thought of as a self-gravitating "gas" of stars. At first sight, the "Milky Way" would seem to be an ideal system on which to test out the ideas of statistical mechanics. Stars in the Galaxy interact via occasional "near miss" events in which they exchange energy and momentum. Actual collisions are very rare indeed. Unfortunately, such interactions take place very infrequently, because there is an awful lot of empty space between stars.

The best estimate for the relaxation time of the "Milky Way" is about 1013 years. This should be compared with the estimated age of the Galaxy, which is only about 1010 years. It is clear that, despite its great age, the "Milky Way" has not been around long enough to reach an equilibrium state. This suggests that the principle of equal a priori probabilities cannot be used to describe stellar dynamics. Not surprisingly, the observed velocity distribution of the stars in the vicinity of the Sun is not governed by this principle.

Reversibility and Irreversibility

Previously, we mentioned that on a microscopic level the laws of physics are invariant under time reversal. In other words, microscopic phenomena look physically plausible when run in reverse. We usually say that these phenomena are reversible. What about macroscopic phenomena? Are they reversible? Well, consider an isolated many particle system which starts off far from equilibrium. According to the H theorem, it will evolve towards equilibrium and, as it does so, the macroscopic quantity H will decrease. But, if we run this process backwards the system will appear to evolve away from equilibrium, and the quantity H will increase.

This type of behaviour is not physical because it violates the H theorem. So, if we saw a film of a macroscopic process we could very easily tell if it was being run backwards. For instance, suppose that by some miracle we were able to move all of the Oxygen molecules in the air in some classroom to one side of the room, and all of the Nitrogen molecules to the opposite side. We would not expect this state to persist for very long. Pretty soon the Oxygen and Nitrogen molecules would start to intermingle, and this process would continue until they were thoroughly mixed together throughout the room.

This, of course, is the equilibrium state for air. In reverse, this process looks crazy! We would start off from perfectly normal air, and suddenly, for no good reason, the Oxygen and Nitrogen molecules would appear to separate and move to opposite sides of the room. This scenario is not impossible, but, from everything we know about the world around us, it is spectacularly unlikely! We conclude, therefore, that macroscopic phenomena are generally irreversible, because they look "wrong" when run in reverse.

How does the irreversibility of macroscopic phenomena arise? It certainly does not come from the fundamental laws of physics, because these laws are all reversible. In the previous example, the Oxygen and Nitrogen molecules got mixed up by continually scattering off one another. Each individual scattering event would look perfectly reasonable viewed in reverse, but when we add them all together we obtain a process which would look stupid run backwards. How can this be? How can we obtain an irreversible process from the combined effects of very many reversible processes? This is a vitally important question. Unfortunately, we are not quite at the stage where we can formulate a convincing answer. Note, however, that the essential irreversibility of macroscopic phenomena is one of the key results of statistical thermodynamics.

Probability Calculations

The principle of equal a priori probabilities is fundamental to all statistical mechanics, and allows a complete description of the properties of macroscopic systems in equilibrium. In principle, statistical mechanics calculations are very simple. Consider a system in equilibrium which is isolated, so that its total energy is known to have a constant value somewhere in the range E to E + ΣE. In order to make statistical predictions, we focus attention on an ensemble of such systems, all of which have their energy in this range. Let Ω(E) be the total number of different states of the system with energies in the specified range. Suppose that among these states there are a number Ω(E;yk) for which some parameter y of the system assumes the discrete value yk. The principle of equal a priori probabilities tells us that all the Ω(E) accessible states of the system are equally likely to occur in the ensemble. It follows that the probability P(yk) that the parameter y of the system assumes the value yk is simply,

$$P(yk) = \frac{\Omega(E; y_k)}{\Omega(E)}$$

Clearly, the mean value of y for the system is given by,

$$y = \frac{\Sigma_k \Omega(E; y_k) y_k}{\Omega(E)}$$

where the sum is over all possible values that y can assume. In the above, it is tacitly assumed that W(E)Σ, which is generally the case in thermodynamic systems.

It can be seen that, using the principle of equal a priori probabilities, all calculations in statistical mechanics reduce to simply counting states, subject to various constraints.

In principle, this is fairly straightforward. In practice, problems arise if the constraints become too complicated. These problems can usually be overcome with a little mathematical ingenuity. Nevertheless, there is no doubt that this type of calculation is far easier than trying to solve the classical equations of motion (or Schrödinger's equation) directly for a many-particle system.

Behaviour of the Density of States

Consider an isolated system in equilibrium whose volume is V, and whose energy lies in the range E to E+ΣE. Let Ω(E,V) be the total number of microscopic

states which satisfy these constraints. It would be useful if we could estimate how this number typically varies with the macroscopic parameters of the system.

The easiest way to do this is to consider a specific example. For instance, an ideal gas made up of spinless monatomic particles. This is a particularly simple example, because for such a gas the particles possess translational but no internal (e.g., vibrational, rotational, or spin) degrees of freedom. By definition, interatomic forces are negligible in an ideal gas. In other words, the individual particles move in an approximately uniform potential. It follows that the energy of the gas is just the total translational kinetic energy of its constituent particles.

Thus,

$$P(yk) = \frac{\Omega(E;y_k)}{\Omega(E)}$$

where m is the particle mass, N the total number of particles, and Pi the vector momentum of the ith particle.

Consider the system in the limit in which the energy E of the gas is much greater than the ground-state energy, so that all of the quantum numbers are large. The classical version of statistical mechanics, in which we divide up phase-space into cells of equal volume, is valid in this limit. The number of states Ω(E,V) lying between the energies E and E + E is simply equal to the number of cells in phase-space contained between these energies. In other words, ΣE, VΣ two energies:

$$Z = \sum_R e^{-\beta(n_1 e_1 + n_2 e_2 + \cdots)}$$

is proportional to the volume of phase-space between these

$$y = \frac{\Sigma_k \Omega(E;y_k)y_k}{\Omega(E)}$$

where (xi yi zi), and (pix piy piz), are the Cartesian coordinates and momentum components of the ith particle, respectively. The integration is over all coordinates and momenta such that the total energy of the system lies between E and E +ΣE. For an ideal gas, the total energy E does not depend on the positions of the particles. This means that the integration over the position vectors ri can be performed immediately. Since each integral over ri extends over the volume of the container (the particles are, of course, not allowed to stray outside the container)

Equation (above) describes the locus of a sphere of radius RE 2mE in the 3N-dimensional space of the momentum components. Hence, X(E) is proportional to the volume of momentum phase-space contained in the spherical shell lying between the sphere of radius R(E) and that of slightly larger radius R(E + 2E) This volume is proportional to the area of the inner

sphere multiplied by 2R=R (E + 2E)–R(E)*. Since the area varies like R3N–1, where B is a constant independent of V or E, and we have also made use of N>>1. Note that, since the number of degrees of freedom of the system is f = 3N, the above relation can be very approximately written,

$$\Omega(E,V)\Sigma \text{ VfEf.}$$

In other words, the density of states varies like the extensive macroscopic parameters of the system raised to the power of the number of degrees of freedom. An extensive parameter is one which scales with the size of the system (e.g., the volume). Since thermodynamic systems generally possess a very large number of degrees of freedom, this result implies that the density of states is an exceptionally rapidly increasing function of the energy and volume. This result, which turns out to be quite general, is very useful in statistical thermodynamics.

❐

6

Entropy in Quantum Mechanics

In quantum statistical mechanics, the concept of entropy was developed by John von Neumann and is generally referred to as "von Neumann entropy". Von Neumann established a rigorous mathematical framework for quantum mechanics with his work Mathematische Grundlagen der Quantenmechanik He provided in this work a theory of measurement, where the usual notion of wave collapse is described as an irreversible process (the so called von Neumann or projective measurement). Using this concept, in conjunction with the density matrix he extended the classical concept of entropy into the quantum domain. It is well known that a Shannon based definition of information entropy leads in the classical case to the Boltzmann entropy.

It is tempting to regard the Von Neumann entropy as the corresponding quantum mechanical definition. But the latter is problematic from quantum information point of view.

Consequently Stotland, Pomeransky, Bachmat and Cohen have introduced a new definition of entropy that reflects the inherent uncertainty of quantum mechanical states. This definition allows to distinguish between the minimum uncertainty entropy of pure states, and the excess statistical entropy of mixtures.

ENTROPY

Concept

The concept of entropy is fundamental to understanding the second law of thermodynamics. Entropy (or more specifically, increase in entropy) is defined as heat (in calories or Btu's) absorbed by a system, divided by the absolute temperature of the system at the time the heat is absorbed. Absolute temperature is the number of degrees above "absolute zero", the coldest temperature that can exist.

The total entropy in a system is represented by the symbol ΔS. The symbol S is used to represent a given change in the entropy content of a system. If the symbol q is used to represent the amount of heat absorbed by a system, the equation for the resulting entropy increase is:

$$\Delta S = q/T \quad (1)$$

Where T is the absolute temperature. When heat is absorbed, the entropy of a system increases; when heat flows out of a system, its entropy decreases.

The "surroundings" of a system is everything outside of the system that can interact with it; surroundings can usually be defined as the space that surrounds a system. When heat is evolved by a system, that same heat is absorbed by its surroundings.

When heat is absorbed by a system, that same heat must necessarily come from its surroundings. Therefore any entropy increase in a system due to heat flow must be accompanied by an entropy decrease in the surroundings, and vice versa. When heat flows spontaneously from a hotter region to a cooler region, the entropy decrease in the hotter region will always be less than the entropy increase in the cooler region, because the greater the absolute temperature, the smaller the entropy change for any particular heat flow.

As an example, consider the entropy change when a large rock at 500 degrees absolute is dropped into water at 650 degrees absolute. (We are using an absolute temperature scale based on Fahrenheit degrees; on this scale, water freezes at 492 degrees.) For each Btu of heat that flows into the rock at these temperatures the entropy increase in the rock is 1/500 = 0.0020 and the entropy decrease of the water is 1/650 = 0.0015. The difference between these values is 0.0020 - 0.0015 = 0.0005. This represents the over all entropy increase of the system (rock) and its surroundings (water).

Of course the rock will warm up to, and the water cool to, a temperature intermediate between their original temperatures, thus considerably complicating the calculation of total entropy change after equilibrium is achieved. Nevertheless, for every Btu of heat transferred from water to rock there will always be an increase of over-all net entropy.

As was mentioned before, a spontaneous change is an irreversible change. Therefore an increase in the overall net entropy can be used as a measure of the irreversibility of spontaneous heat flow.

Irreversible changes in a system can, and often do, take place even though there may be no interaction, and negligible heat flow, between system and surroundings. In cases like these the entropy "content" of the system is greater after the change than before. Even when heat flow does not occur between system and surroundings, spontaneous changes inside an isolated system are always accompanied by an increase in the system's entropy, and this calculated entropy increase can be used as a measure of irreversibility. The following paragraphs will explain how this entropy increase can, at least in some cases, be calculated.

It is an axiom of thermodynamics that entropy, like temperature, pressure, density, etc., is a property of a system and depends only on the existing condition of the system. Regardless of the procedures followed in achieving a given condition, the entropy content for that condition is always the same. In other words, for any given set of values for pressure, temperature, density, composition, etc., there can be only one value for the entropy content. It is essential to remember this: When a system that has undergone an irreversible change is restored to its original condition (same temperature, pressure, volume, etc.) its entropy content will likewise be the same as it was before.

In cases where an isolated system undergoes an entropy increase as the result of a spontaneous change inside the system, we can calculate that entropy increase by postulating a procedure whereby the system's entropy increase is transferred to the surroundings in a manner such that there is no further increase in net entropy and the system is restored to its original condition. The entropy increase of the surroundings can then be readily calculated by equation:

(1) $DS = q/T$, where q = heat absorbed by the surroundings, and T = absolute temperature of the surroundings.

It bears repeating that when the system is restored to its original condition, its entropy content will be the same as it was before its irreversible change. Therefore the amount of entropy absorbed by the surroundings during restoration must necessarily be the same as the entropy increase accompanying the system's original irreversible change, providing that there is no further increase in net entropy during restoration.

This postulated restoration procedure and the postulated properties of the surroundings are for the purpose of calculation only. Since we are not dealing with the surroundings as such, they can be postulated in whatever form necessary to simplify the calculations; it is neither necessary nor desirable that the surroundings correspond to any condition that could actually exist. Therefore, we will postulate a theoretical restoration procedure that takes place with no further increase in net entropy, even though such a procedure can not actually be obtained experimentally.

The restoration process, if it were to take place in actuality, would have to be accompanied by at least a small amount of irreversibility, and hence an additional increase in the entropy of the surroundings beyond the entropy increase from the system's original irreversible change. This is because heat will not flow without a temperature differential, friction cannot be entirely eliminated, etc. Therefore the restoring process, if it is to take place with no further increase in over-all net entropy, must be postulated to take place with no irreversibility.

If such a process could be actually realized, it would be characterized by a continuous state of equilibrium (i.e. no pressure or temperature differentials) and would occur at a rate so slow as to require infinite time. Processes like these are called "reversible" processes. Remember, reversible processes are postulated to simplify the calculation of the entropy change in a system; it is not necessary that they be capable of being achieved experimentally.

It should not be assumed that equation requires that q, the heat absorbed, must necessarily be absorbed reversibly. The concept of reversibility is merely a means to an end: the calculation of entropy change accompanying an irreversible process.

The following example will illustrate the calculation of a reversible restoring process and at the same time develop the equation which is the basis for the thermodynamical relationship between probability and the second law. We will postulate a system consisting of an "ideal" gas contained in a tank connected to a second tank that has been completely evacuated, with the valve between the two tanks closed. The temperature of the system and its surroundings is postulated to be the same. An ideal gas is one whose molecules are infinitely small and have no attractive or repulsive forces on each other. (Under ordinary conditions hydrogen and helium closely approximate the properties of an ideal gas.) An ideal gas is chosen in order to develop the basic relationship without introducing complicating correction factors to account for the size of the molecules and the forces they exert on each other. When the valve is opened the gas expands irreversibly from V1 (its original volume) to V2 (the volume of both tanks). There is no work of compression by or upon the surroundings. Because the gas is ideal there is no temperature change, and hence no heat flow takes place.

After expanding irreversibly from V1 to V2, the gas is restored to its original condition by reversibly compressing it back to V1. This compression requires work (force applied through a distance) which in turn generates heat in the gas, heat that is absorbed by the surroundings so that there is no increase in the gas temperature. In our mathematical model of this reversible restoring process, the surroundings are postulated to be so large that they also do not undergo any temperature increase. The temperature T remains unchanged during the entire irreversible expansion and subsequent reversible restoration process. The work of compressing the gas during restoration is equal to the pressure of the gas times the volume change due to compression. Because the pressure increases during compression, the work of compression must be determined by the calculus integral:

$$\text{Compression Work} = \text{PdV}$$

where: P = pressure

V = volume

dV = the small change in volume taking place at the corresponding pressure P

The integral sign indicates the summation of all the individual values of PdV.

The equation relating temperature, pressure, and volume of an ideal gas is:

$$PV = RT \quad (2)$$

where: P = pressure

V = volume

T = absolute temperature

R = a constant which depends only on the amount of gas present

In the case of a reversible, isothermal compression of an ideal gas we may substitute P from equation (2) into the equation for compression work.

When this is done, we have:

$$\text{Compression Work} = RT\ dV/V \quad (3)$$

Although it is not necessary that our postulated reversible restoration process be capable of being carried out in a practical sense, it is nevertheless sometimes helpful to be able to visualize the process. To this end, the reader may consider the restoring compression process being brought about by a piston fitted into the end of the second tank. On compression from V2 to V1, the piston moves down the length of the second tank, and with no mechanical friction forces all the gas contained therein back into the first tank V1.

Since the work of compression is equal to q, the heat absorbed by the surroundings, q may be substituted in equation (3) to give:

$$q = RT\, dV/V \quad (4)$$

From equation (1) the entropy gained by the surroundings during restoration from V2 to V1 is:

$$\Delta S = q/T \quad (1)$$

Substituting from equation (4):

$$\Delta S = RdV/V$$

Upon integrating (a calculus procedure for summing up the individual values of dV/V) we have:

$$\Delta S = R\ln(V2/V1) \quad (5)$$

Where in (V2/V1) is the natural logarithm of the ratio of expanded volume to the initial volume, and ΔS is equal to the entropy increase in the surroundings upon restoration compression from V2 to V1. As we have seen, ΔS is also equal to the entropy increase of the gas caused by its original expansion from V1 to V2. This is because V1 is the same volume both before expansion and after restoration compression, and therefore has the same entropy content. Therefore 206 Statistical Mechanics the entropy transferred to the surroundings during restoration is equal to that gained by the system in expanding from V1 to V2.

ENTROPY AND PROBABILITY

The ratio of the probability that all the gas molecules are evenly distributed between the two tanks to the probability that all the molecules, of their own accord and by random motion, would be in tank V1 is equal to (V2/V1)N, where N is the number of molecules. If V2/V1 were equal to 2.0, for example, and N were equal to 10, the probability ratio would be 2 to the tenth power, or 1024. For N = 100, the ratio would be approximately 1.27 times ten to the 30th power. It is clear that the random motion of trillions of gas molecules heavily favours a uniform distribution.

let X1 = the probability of all the gas molecules, after the valve is opened, remaining in the first tank V1

let X2 = the probability of all the gas molecules, after the valve is opened, being uniformly distributed in V2, the volume of both tanks.

From the probability equation, we have:

$$X2/X1 = (V2/V1)^{N}.$$

Taking the natural logarithm of both sides, and then multiplying both sides by R, the gas constant:

$$R \ln(X2/X1) = RN \ln(V2/V1)$$

$$R/N \ln(X2/X1) = R \ln(V2/V1)$$

Substituting in equation (5):

$$\Delta S = R/N \ln(X2/X1) \quad (6)$$

Equation (6) represents the fundamental relationship between probability and the second law of thermodynamics. It states that the entropy of a gaseous system increases when its molecular distribution changes from a lower probability to a higher probability (X2 greater than X1).

Based on the belief that the laws of thermodynamics are universal, this equation has been assumed to apply to all systems, not just gaseous. In other words, any entropy change is proportional to the logarithm of the ratio of probabilities. Therefore, for the general case equation (6) can be written:

$$\Delta S = K \ln(X2/X1) \quad (7)$$

Where K is a constant depending on the particular change involved. However, individual values of K, X1, or X2 are seldom, if ever, known for non-gaseous systems.

As we have seen before, ΔS can be either positive or negative.

When ΔS is negative equation (7) can be written:

$$-\Delta S = -K \ln(X2/X1)$$

$$= K \ln(X1/X2)$$

Therefore a system can go from a more probable state (X2) to a less probable state (X1), providing S for the system is negative. In cases where the system interacts with its surroundings, S can be negative providing the over-all entropy of the system and its interacting surroundings is positive; the over-all change can be positive if the entropy increase of the surroundings is numerically greater than the entropy decrease of the system. In the case of the formation of the complex molecules characteristic of living organisms, creationists raise the point that when living things decay after death, the process of decay takes place with an increase in entropy.

They also point out, correctly, that a spontaneous change in a system takes place with a high degree of probability. They fail to realize, however, that probability is relative, and a spontaneous change in a system can be reversed, providing the system interacts with its surroundings in such a manner that the entropy increase in the surroundings is more than enough to reverse the system's original entropy increase.

The application of energy can reverse a spontaneous, thermodynamically "irreversible" reaction. Leaves will spontaneously burn (combine with oxygen) to form water and carbon dioxide. The sun's energy, through the process of photosynthesis, will produce leaves from water vapour and carbon dioxide, and form oxygen.

If we unplug a refrigerator, heat will flow to the interior from the surroundings; the entropy increase inside the refrigerator will be greater than the entropy decrease in the surroundings, and the net entropy change is positive. If we plug it in, this spontaneous "irreversible" change is reversed. Due to the input of electrical energy to the compressor, the heat transferred to the surroundings from the condenser coils is greater than the heat extracted from the refrigerator, and the entropy increase of the surroundings is greater than the entropy decrease of the interior, in spite of the fact that the surroundings are at a higher temperature. Here again, the net entropy change is positive, as would be expected for any spontaneous process.

In a similar manner, the application of electrical energy can reverse the spontaneous reaction of hydrogen and oxygen to form water: when a current is passed through a water solution, hydrogen is liberated at one electrode, oxygen at the other.

As can easily be confirmed experimentally, agitating water raises its temperature. When water falls freely from a higher elevation to a lower elevation, its energy is changed from potential to kinetic, and finally to heat as it splashes at the end of its fall. The second law of thermodynamics states that the water will not spontaneously raise itself to its original elevation using the heat produced on splashing as the sole source of energy. To do so would require a heat engine that would convert all of the heat of splashing to mechanical energy.

The efficiency of a heat engine is thermodynamically limited by the Carnot cycle, which limits the efficiency of any heat engine to $\Delta T/T$, where ΔT is the temperature increase due to splashing, and T is the absolute temperature. Since ΔT is only a small fraction of T, there is no device that could be constructed which would allow all of the water to spontaneously jump back to its former elevation.

We can, at least in theory, calculate the entropy increase of the water resulting from its irreversible change in falling. In a manner analogous to that used in the previous example, the entropy increase would be equal to the heat generated by splashing agitation, divided by the absolute temperature. If some of the energy of the falling water is extracted by a water wheel, there will be less heat of splashing and hence less entropy increase.

A properly designed turbine could extract most of the water's kinetic energy. This is not the same thing as trying to utilize the heat of splashing as an energy source for a heat engine to raise the water. In other words, using the energy before it becomes heat is much more efficient than trying to use it after it becomes heat. If a water wheel is connected by shafts, belts, pulleys, etc. to a pump, the pump can raise water from the downstream side of the water wheel to an elevation even higher than that of the upstream reservoir. Some of the water would spontaneously raise itself to an elevation even higher than original, but the rest of it would end up below the water wheel on the downstream side. While it is not possible for all of the water to raise itself to an elevation higher than its initial elevation, it is possible for some of the water to spontaneously raise itself to an elevation higher than initial. As with any other irreversible change, there will be an increase in over-all entropy. This means that the entropy increase of the water going over the wheel is greater than the entropy decrease of water pumped up to the higher elevation.

This will be confirmed mathematically in the following paragraphs. will stand for the Greek letter gamma, representing unit weight in pounds per cubic foot. An increase in the value of a parametre will be represented by Δ.

Let:

γ = unit weight of water, lbs./cubic foot

Δh = height of reservoir above downstream side, feet

γh = potential energy of water in reservoir

Δh = additional height above reservoir (to which water is pumped)

$h + \Delta h$ = height to which water is pumped, feet above downstream side

w = work of pumping water to higher elevation

q = heat loss due to pump friction and downstream agitation

f = fraction of water pumped to elevation h + h

From the flow equation, energy in = energy out:

$$\gamma h = q + w$$

Where:
h = the total energy expended by the water in falling from height h.
q = the energy wasted as heat of splashing
w = work expended in pumping the water to a higher elevation
q = TΔS (from equation 1)

$$w = f\gamma(h + \Delta h)$$

The total energy available, γh, is divided into pump work, fγ(h + Δh), and energy
lost, TΔS:

$$\gamma h = T\Delta S + f\gamma(h + \Delta h)$$

Rearranging:

$$T\Delta S = \gamma h - f\gamma(h + \Delta h) \quad (8)$$

When no pump work is done, then:

$$T\Delta S' = \gamma h \quad (9)$$

Combining equations (8) and (9), we get:

$$\Delta S' - \Delta S = f\gamma(h + \Delta h)/T \quad (10)$$

In the case where the water falls freely without turning the water wheel or operating the pump:
let q′= heat of splashing

$$q' = T\Delta S'$$

where ΔS′ is the entropy increase on free fall

Equation (10) shows that ΔS′ is larger than ΔS, and that the entropy increase due to pump friction and downstream agitation is "backed up" by the even larger entropy increase that takes place when water falls freely. Equation (10) also shows that the lower the value of ΔS, the more efficient the pump, and the greater the value of f, the fraction of water pumped.

Creationists assume that a change characterized by a decrease in entropy can not occur under any circumstances. In fact, spontaneous entropy decreases can, and do, occur all the time, providing sufficient energy is available. The fact that the water wheel and pump are man-built contraptions has no bearing on the case: thermodynamics does not concern itself with the detailed description of a system; it deals only with the relationship between initial and final states of a given system (in this case, the water wheel and pump).

A favourite argument of creationists is that the probability of evolution occurring is about the same as the probability that a tornado blowing through a junkyard could form an airplane. They base this argument on their belief that changes in living things have a very low probability and could not occur without "intelligent design" which overcomes the laws of thermodynamics. This represents a fundamental contradiction in which (they say) evolution is inconsistent with thermodynamics because thermodynamics doesn't permit order to spontaneously arise from disorder, but creationism (in the guise of intelligent design) doesn't have to be consistent with the laws of thermodynamics.

A simpler analogy to the airplane/junkyard scenario would be the stacking of three blocks neatly on top of each other. To do this, intelligent design is required, but stacking does not violate the laws of thermodynamics.

The same relations hold for this activity as for any other activity involving thermodynamical energy changes. It is true that the blocks will not stack themselves, but as far as thermodynamics is concerned, all that is required is the energy to pick them up and place them one on top of the other.

Thermodynamics merely correlates the energy relationships in going from state A to state B. If the energy relationships permit, the change may occur. If they don't permit it, the change can not occur.

A ball will not spontaneously leap up from the floor, but if it is dropped, it will spontaneously bounce up from the floor. Whether the ball is lifted by intelligent design or just happens to fall makes no difference. On the other hand, thermodynamics does not rule out the possibility of intelligent design; it is just simply not a factor with respect to the calculation of thermodynamic probability. Considering the earth as a system, any change that is accompanied by an entropy decrease (and hence going back from higher probability to lower probability) is possible as long as sufficient energy is available. The ultimate source of most of that energy, is of course, the sun.

The numerical calculation of entropy changes accompanying physical and chemical changes are very well understood and are the basis of the mathematical determination of free energy, emf characteristics of voltaic cells, equilibrium constants, refrigeration cycles, steam turbine operating parametres, and a host of other parametres. The creationist position would necessarily discard the entire mathematical framework of thermodynamics and would provide no basis for the engineering design of turbines, refrigeration units, industrial pumps, etc. It would do away with the well-developed mathematical relationships of physical chemistry, including the effect of temperature and pressure on equilibrium constants and phase changes.

PHYSICAL CONCEPT OF ENTROPY

Conservation of energy

The conservation of energy is the bedrock of our understanding of nature. Conservation of energy, however, only tells us that in any process the total energy remains constant, that is . Energy conservation, however, does not specify what can enter and leave the energy conservation equation, only that the sum of energy entering and leaving a system must always be the same. Furthermore, energy conservation does not tell us what are the various forms of energy, and how energy transforms from one form into another. From our study of waves, light, electric and magnetic fields and so on, we found that energy plays a central role in all these phenomena.

However, only the specific study of the various phenomena could unravel the specific forms that energy takes. The concept of energy also does not tell us what fraction of the total energy is kinetic and what fraction is electrical, chemical and so on. A new concept,in addition to energy, is needed for a more complete understanding of physical processes, in particular those involving the transformation of energy. In all the examples we studied, energy could be completely transformed from one form into another.

For example, in our study of magnetic fields, we found that we could fully convert mechanical energy into electrical currents by Faraday's Law of Induction.

We may well ask: can all forms of energy be fully transformed from one form to another? The answer is unexpectedly complicated. In some cases the answer is "yes", and in other cases the answer is "no". The concept of entropyaddresses this question regarding the transformation of the forms of energy, and tells us how much energy can be converted from one form into another, and in particular, how much energy is available for doing useful work. Heat had defied all attempts to explain it using the concept of force, and, in general using Newtonian mechanics.

The fact that heat is a form of energy presented a great breakthrough in the understanding of heat, and was fundamental to the emergence of energy as one of the most fundamental ideas in physics. It was, however, soon realized that heat was not an ordinary form of energy such as gravitational or electrical energy.

Rather, heat is a unique form of energy in that only a certain fraction of heat energy can be transformed into other forms of energy. It is for understanding the unique problems presented by heat that new concepts such as entropy, which go beyond energy, were developed. The idea of entropy was introduced into physics in the nineteenth century and is essential for understanding the phenomenon of heat. Entropy is derived from the Greek word meaning "transformation content".

Famous names such as Sadi Carnot, R.J. Clausius, Count Rumford, James Joule, Ludwig Boltzmann and so on contributed greatly to the ideas of heat and entropy. Entropy is a concept which does not answer all the questions left unanswered by energy, but it does address a crucial aspect of energy, namely how much energy can be transformed from one form into another. In particular, if a certain amount of heat energy is given, then entropy tells us how much "useful" work can be extracted from heat energy, in other words, what is the maximum possible efficiency of a heat engine.

In general, since energy is conserved, one would think there is no need to be concerned about recklessly wasting energy. But we know from daily life that "useful" energy, or equivalently, energy "available" for use, is a scarce resource. This intuitive understanding that useful energy is a precious resource is explained by the concept of entropy. Entropy is also related to the concept or order and disorder, and a phase transition from say water to gas is permissible only if entropy increases in such a transformation.

For any system, entropy is a physically measurable quantity. Furthermore, it is an experimentally observed fact that for all processes in nature, the entropy of the total system can never decrease. Note this empirical law concerning entropy is weaker than that of energy, since it does not specify, for any physical process, by how much does entropy increase.

The concept of entropy has many other applications and vast ramifications in other disciplines. Only those chemical processes and reactions for which the total entropy of the system does not decrease are allowed by nature. All living entities must maintain their low entropic state to stay alive. More abstract applications of entropy occur in the physics of black holes, and it was only in 1997 that the entropy of black holes could be calculated from first principles using results from string theory.

The concept of entropy also has a key application in information theory. It was shown by Claude Shannon in 1950 that the information content of a message is determined by a suitable application of the idea of entropy to the discipline of information science. The current understanding of entropy is that it is a large scale (macroscopic) manifestation of the atomic nature of matter.

For example, one cubic centimetre of air contains about number of atoms. This is an unimaginably large number of particles.

Even if under some circumstances we can view these atoms as classical particles, it is futile to apply Newton's laws to such a large collection of particles. Instead, the best we can do is to acknowledge our total ignorance and assume that all the atoms are moving randomly, that is, the position and velocity of every atom is a random variable. Furthermore, for a system of atoms that reaches a stable state, called equilibrium (more on this concept later), every possible velocity and position of the atoms in one cubic centimetre of air is equally likely. \We will show later that the empirical law that entropy never decreases is simply a statement that for a system consisting of a collection of atoms which is not in equilibrium, the collection of atoms will always move towards a state that is the more probable, and will only reach equilibrium on arriving at the most probable state available to the system.

It should be noted that the field of thermodynamics developed in the eighteenth and nineteenth centuries without the concept of the atomic composition of matter.

It is remarkable that very general results on how bulk matter behaves were obtained without any idea of the microscopic composition of matter.

From the point of view of scientific methodology, the progress of thermodynamics shows that there are concepts describing the large scale properties of matter that have a logic of their own, and which later on are seen to match on smoothly to a deeper and more complete understanding of the same phenomenon.

Temperature and Heat

We discuss thermodynamics, which is the branch of physics dealing with heat, because this will give us a specific physical system in which we can precisely define and discuss entropy and related ideas. In particular, we will study the behaviour of of an ideal gas to illustrate the concept of heat, energy and entropy.

Most of us have an intuitive understanding of heat, based on having experienced hot and cold objects. The first thing one must realize is that heat is different from temperature.

ENTROPY IN ASTROPHYSICS

In astrophysics, what is referred to as "entropy" is actually the adiabatic constant derived as follows. Using the first law of thermodynamics for a quasistatic, infinitesimal process for a hydrostatic system,

$$dQ = dU = dW.$$

For an ideal gas in this special case, the internal energy, U, is only a function of T; therefore the partial derivative of heat capacity with respect to T is identically the same as the full derivative, yielding through some manipulation,

$$dQ = CV\ dT = PdV.$$

Further manipulation using the differential version of the ideal gas law, the previous equation, and assuming constant pressure, one finds,

$$dQ = CPdT = VdP.$$

One can solve this simple differential equation to find,

$$PV\gamma = \text{constant} = K$$

where k_B is Boltzmann's constant. Substituting this into the above equation along with V = [grams]/γ and γ = 5/3 for an ideal monoatomic gas one finds,

$$K = \frac{k_B T}{\mu m_H r^{2/3}},$$

where μ is the mean molecular weight of the gas or plasma; and μH is the mass of the Hydrogen atom, which is extremely close to the mass of the proton, mp, the quantity more often used in astrophysical theory of galaxy clusters. This is what astrophysicists refer to as "entropy" and has units of [keV cm2]. This quantity relates to the thermodynamic entropy as,

$$S = k_B \ln \Omega + S0$$

where Ω, the density of states in statistical theory, takes on the value of K as defined above.

Energy Dispersal

The concept of entropy can be described qualitatively as a measure of energy dispersal at a specific temperature. Similar terms have been in use from early in the history of classical thermodynamics, and with the development of statistical thermodynamics and quantum theory, entropy changes have been described in terms of the mixing or "spreading" of the total energy of each constituent of a system over its particular quantized energy levels. Ambiguities in the terms disorder and chaos, which usually have meanings directly opposed to equilibrium, contribute to widespread confusion and hamper comprehension of entropy for most students. As the second law of thermodynamics shows, in an isolated system internal portions at different temperatures will tend to adjust to a single uniform temperature and thus produce equilibrium.

A recently developed educational approach avoids ambiguous terms and describes such spreading out of energy as dispersal, which leads to loss of the differentials required for work even though the total energy remains constant in accordance with the first law of thermodynamics. Physical chemist Peter Atkins, for example, who previously wrote of dispersal leading to a disordered state, now writes that "spontaneous changes are always accompanied by a dispersal of energy", and has discarded 'disorder' as a description.

ENTROPY CHANGE FOR IDEALLY CLOSED SYSTEM GAS

For an ideal constant volume process... ($dQ = mc_v dT$)

$$dS = \frac{dQ}{T} = \frac{mc_v dT}{T}$$

Integrating between initial (2) and final state (2)

$$S_2 - S_1 = mc_v \log_\theta \frac{T}{T_1}$$

For an ideal constant pressure process... (dQ = mcpdT)

$$dS = \frac{dQ}{T} = \frac{mc_p dT}{T}$$

Integrating between initial (2) and final state (2)

$$S_2 - S_1 = mc_p \log_e \frac{T_2}{T_1}$$

For an ideal adiabetic process...

$$dS = \frac{dQ}{T} = 0$$

For an ideal isothermal process...

There is no change in temperature and therefore there is no change in internal energy U.

From the first law of thermodynamic dQ = dU + dW. If dU = 0 then dQ = dW.

$$dS = \frac{dW}{T} = \frac{pdV}{T}$$

$$aspV = mRT \; or \; T = \frac{pV}{mR}$$

$$dS = \frac{pdVmR}{pV} = \frac{mRdV}{V}$$

Integrating between initial (2) and final state (2)

$$S_2 - S_1 = mR\log_e \frac{V_2}{V_1}$$

Entropy changes in terms of properties of a perfect gas

From the definition of entropy i.e δQrev = TδS and from the first law of thermodynamics is dQ = dE + dW which, for a reversible process can be written as δQrev =δE + PδV... it can be deduced that dE = TdS - PdV for a reversible process and if the substance is a gas of mass m then

dQ = TdS = mcvdT +PdV........equationX

For a perfect gas PV = mRT and assuming Cv is constant.

THE GENERIC BALANCE EQUATION

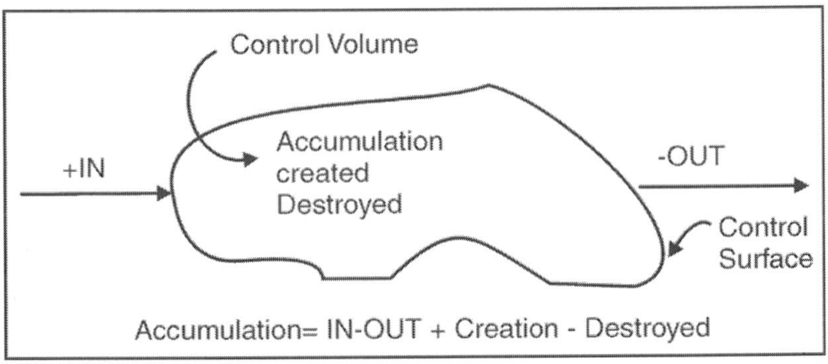

The Generic Balance Equation

The generic balance equation is so simple that, if we described it incorrectly to a three-year-old child, he or she would recognize that something was wrong, which is a good point in favour of the position that reasonableness is innate, i.e., an a priorisynthetic judgment. One may argue as to what this equation may be applied to; but, if we should claim that it applies to what is commonly known as stuff, your objection would be a mere quibble. [Under the aegis of the commonality of the word stuff we consider both corporeal and incorporeal elements.

For example, energy is conserved, however it may be more like the behaviour of a mysterious something than the something that behaves. It may be so abstract that nothing but a mathematical mapping of the relevant portion of the Universe is adequate to describe it. Mathematicians are content to refer to this mapping as field equations. As far as they are concerned, the field equations are the phenomenon.

The generic balance (or accounting) equation states quite simply that the accumulation within the control volume equals whatever is created within the control volume minus whatever is destroyed inside the control volume plus whatever enters the control volume minus whatever leaves the control volume. The situation is illustrated in Figure. By the accumulation we mean the difference between what we ended up with and what we started out with. This can be negative or positive; but, if it be negative, we might call its absolute value the deficit. As stated above, the control volume is any well-defined region in space. It may be changing shape and moving and it need not be a connected set.

Many purists will object that the balance equations are not the laws of thermodynamics, which, according to natural philosophy, must be statements that come entirely from experience and may not employ such abstract concepts as energy, temperature, and entropy. In particular, the Second Law should be a statement about a particular type of physical device that cannot exist. In fact, we have two statements each with its own impossible device. We suggest that the reader consult the excellent book for the layman by P.W. Atkins. This will not be our last mention of this book. Suffice it to say that the balanceequation approach is logically, if not philosophically, equivalent to the experiential statements of the laws.]

We now wish to describe the balance equations of thermodynamics, which we are giving the status of laws, the famous laws of thermodynamics. We shall describe the First Law first. Since the First Law is an energy balance, since everyone thinks he knows what energy is, and since entropy, which is an important property of thermodynamic systems, does not appear in the First Law, most thermodynamics texts do not define entropy before they discuss the First Law of Thermodynamics, which involves both work and heat. Work and heat are not properties of the system, but they are rather subtle concepts.

Under some circumstances work and heat are mistaken for one another; whereas, if they are defined in terms of entropy, they can be distinguished easily – at least from the theoretical point of view. Therefore, we shall define and discuss entropy at this time. In our opinion, teachers of thermodynamics should give a little consideration to this departure from the usual way of presenting the First Law.

CHANGE IN ENTROPY

Entropy is used in everyday life as a synonym for chaos. For example: the entropy in my room increases as the semester goes on. But it's also been used to describe the approach to an imagined final state of the universe when everything reaches the same temperature: the entropy is supposed to increase to a maximum, then nothing will ever happen again.

Of course, the first law—the conservation of total energy including heat energy—is easy to express quantitatively: one only needs to find the equivalence factor between heat units and energy units, calories to joules, since all the other types of energy are already in joules, add it all up to get the total and that will remain constant.

The second law, that heat only flows from a warmer body to a colder one, does have quantitative consequences: the efficiency of any reversible engine has to equal that of the Carnot cycle, and any nonreversible engine has less efficiency.

Entropy (S) is very difficult to visualize because it does not represent anything tangible. The entropy increase δS is the heat transfer to a substance δQ divided by the absolute temperature of the substance (T) during a reversible heat-transfer process. At the very simplest level, on a plot of Absolute Temperature (T) against Entropy ($Q/T = S$) for a reversible cyclic process (as shown below) the area enclosed = Q

Note: The reversible cyclic process shown below is actually the theoretical Carnot cycle.

1->2 being isothermal expansion.
2->3 being adiabetic expansion.
3->4 being isothermal compression.
4->1 being adiabetic compression.

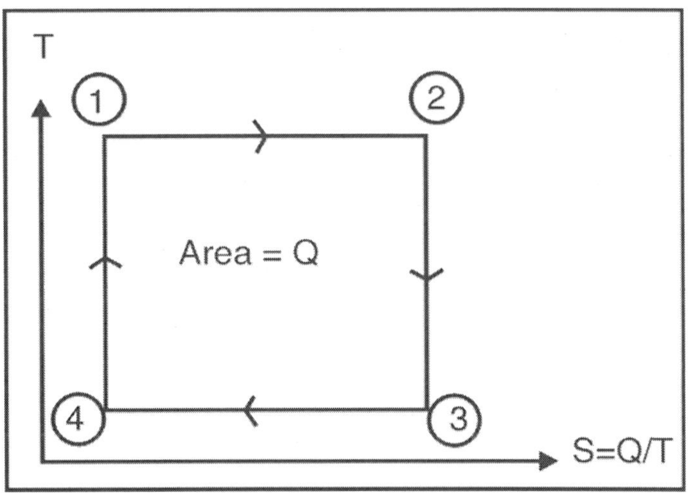

The change in entropy (δS) of a substance is that quantity which when multiplied by the absolute temperature at which the change took place results in the amount of energy (δQ) flow reversibly by heat transfer across the boundary enclosing the substance.

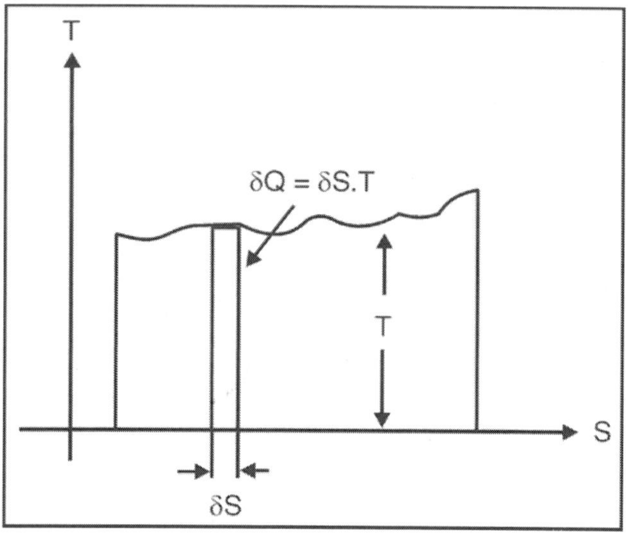

Total entropy is a property of a substance and therefore the change in entropy during a process, from an initial to a final state, is the same whatever the path taken. Therefore to determine the change in entropy resulting from a real irreversible process an equivalent reversible process must be envisaged to replace the real process from initial to final states before integration of the following

$$\int_i^f \frac{dQ_{rQv}}{T} = S_f - S_i$$

Typical SI units for total entropy (S) change are kJ/K. SI units for specific entropy (s) change are kJ/kg.K

Normally, students have a problem with the concept of entropy. This is expected, but what is unfortunate is that they think they understand energy. One supposes that if we use a term enough we think we understand it. After the population crisis and very much related to it, the most serious crisis facing humanity is usually referred to as the energy crisis – even by the President of the United States, high government officials, and famous professors.

Of course it should be referred to as the entropy crisis, availability crisis, or emergy crisis, but we are getting a little ahead of ourselves. (It came as quite a shock to one of us when he discovered that the people who are running the world don't know what they are talking about – much less what they are doing.) To facilitate the definition of entropy without recourse to the Second Law, we shall depart from so-called classical thermodynamics, which forbids inquiry into the microscopic nature of the universe. Just for a moment, we shall take a quick peek at statistical thermomechanics.

Typically, a system the entropy of which we wish to know has been defined in terms of common macroscopic thermodynamic properties that we have at our disposal and with which the reader may already be familiar such as volume, pressure, temperature, and internal energy for which we have not given formal definitions. In addition, let us suppose that the analyst is in possession (hypothetically) of a number of probability distributions Dk ={pi, i = 1, 2, …, Nk} each of which, k = 1..., M, (i) provides the probability that the system will be found in the i-th microscopic state, and (ii) is entirely consistent with the knownmacroscopic properties of the system.

Remember, in classical thermodynamics, we do not inquire deeply into the microscopic picture; therefore; we should not be surprised to find that more than one – perhaps many – such probability distributions could represent precisely the same state viewed macroscopically.

For each such distribution a candidate Sk can be calculated for the entropy of the system.

It is the minimal amount of information – measured in bits, say – to determine from the distribution under investigation the exact microscopic state of the system. It should not be construed that this determination could actually be carried out – even theoretically; but, it is easy to determine the expected (in the probabilistic sense) information deficit corresponding to the known macroscopic thermodynamic variables and the j-th probability distribution,

$$D_j = \{p_i \log p_i i = 1, N_j\} = \left\{ \sum_{i=1}^{N_j} p_i \log_2 (p_i) \right\}_j$$

where Nj different possible and compatible microscopic states are associated with Dj . The macroscopic entropy of the system, S, is the maximum value of the minimal information deficits, i.e., S = maximum{Sj , j = 1 to M}, where M probability distributions are compatible with the macroscopic state of the system as described by classical thermodynamics.

As a somewhat challenging exercise, the reader may show, by considering the old TV game Twenty Questions, that the minimal information to determine the exact microscopic state of the i-th microstate is the log to the base two of pi. This will give the entropy as the amount of information needed in bits, which would be converted into standard thermodynamic units by multiplying the same amount of information expressed as a natural logarithm,

$$\log_2 p_i = \frac{In\ p_i}{In\ 2} = In\left(h2\sqrt{p_i}\right)$$

by Boltzmann's constant, k = Ro/A, where Ro is the universal gas constant and A is Avogadro's number. (Don't worry if you don't know what these numbers are.) The standard units of entropy are energy over temperature – essentially for traditional reasons. Despite the unfortuitous happenstance that Joules per Kelvin is not particularly suggestive of information, entropy is a measure of the quantity of information that would be needed to determine the microstate from the classical thermodynamic macrostate (except for a constant factor) although it is sometimes referred to as a measure of uncertainty, disorder, randomness, or chaos, which, from our view, are less satisfactory interpretations. The preceding remarks on entropy are derived from Dr. David Bowman's generous postings to a list server for physics teachers on the Internet.

Let us continue our mathematical analysis for a simple and commonly encountered probability distribution. Suppose a system be capable of taking [capital omega] different quantum states with the probability for the i-th state being pi. Then, the entropy of that system is S=-k S pi ln pi , where S represents the summation from i = 1 to i = Ω, ln is the natural logarithm [ln pi is the exponent to which the transcendental number e (equal to approximately 2.718282) must be raised to get the number pi] and k is Boltzmann's constant. The case where the probability of each quantum state is the same, namely, when pi = 1/ Ω, is exceptionally famous. In that case, we get S = k ln Ω. The expression S = k ln Ω appears on Boltzmann's tomb – unless we have been hoodwinked by the scientific historians, which is not entirely out of the question, although Truesdell provides a photograph.

We may now distinguish between heat and work, both of which entail the transfer of energy at the boundary and only at the boundary of a control volume; however, heat carries entropy along with it and affects the entropy balance accordingly. Work, on the other hand, carries no entropy, and, therefore, has no effect on the entropy balance. Because of this difference between heat and work an asymmetry arises; namely, work can be converted entirely to heat in every case including the case where no other change occurs in the universe, but heat can be converted completely to work only when the rest of the universe is changed in some additional fundamental way. If we restrict ourselves to cyclic processes, work can be converted completely to heat but not vice versa.

This is easy to visualize: Case 1: Consider the spring escapement in your Grandfather's watch. It transfers work to the gears, pointers, and whatever else accepts the work done by the spring mechanism, which we will take as "the system". The state of the material objects by means of which this energy is transferred as work is completely organized. The microstate is completely determined by the extent to which the spring has become unwound.

The probability of being in that state is one and the log of one is zero. The entropy is zero. Case 2: Imagine, for a moment, a hydraulic watch, say, driven by a pumped hydraulic fluid, which is permitted to flow across the control surface as we conceive it. This appears to be work too, but the turbulence and random motion of the pumped fluid as well as the friction losses in the pipe convert the electricity that drives the pump (work) into part work and part heat due to the turbulence and other forms of fluid friction. Since heat is crossing the control surface it will be accompanied by entropy since clearly the fluid will not be found in a state clearly defined by one parameter that takes its unique value with probability one. Case 3: Finally, if we had a steam clock, we could drive it by heating water in a boiler the surface of which facing the fire is our conceptual control surface. This is a case of heat and only heat crossing the control surface. Consider the entropy associated with this heat. Hint: Imagine how complicated fire is. The complications in the nature of the fire will complicate the conduction of heat to the boiler.]

Some analysts may quibble that heat doesn't cross the control surface. Rather, thermal energycrossing the control surface is heat; that is, crossing the control surface is part of the description of heat not separate from it. We do not feel the necessity for this kind of precision.

PROPERTIES OF ENTROPY

Entropy, as we have defined it, has some dependence on the resolution δE to which the energy of macrostates is measured. Recall that $\Omega(E)$ is the number of accessible microstates with energy in the range E to $E + \delta E$. Suppose that we choose a new resolution δ^*E and define a new density of states Ω^* E which is the number of states with energy in the range E to $E + \delta^*E$. It can easily be seen that,

$$\Omega^*(E) = \frac{\delta^* E}{\delta E}\Omega(E)$$

It follows that the new entropy S*= k ln Ω* is related to the previous entropy S* = k ln Ω via,

$$S^* = S + k\ln\frac{\delta^* E}{\delta E}$$

Now, our usual estimate that Ω ~ Ef gives S ~ kf, where f is the number of degrees of freedom. It follows that even if δ*E were to differ from δE by of order f (i.e., twenty four orders of magnitude), which is virtually inconceivable, the second term on the right-hand side of the above equation is still only of order k ln f, which is utterly negligible compared to kf. It follows that,

$$S^* = S$$

to an excellent approximation, so our definition of entropy is completely insensitive to the resolution to which we measure energy (or any other macroscopic parameter).

Note that, like the temperature, the entropy of a macrostate is only welldefined if the macrostate is in equilibrium. The crucial point is that it only makes sense to talk about the number of accessible states if the systems in the ensemble are given sufficient time to thoroughly explore all of the possible microstates consistent with the known macroscopic constraints. In other words, we can only be sure that a given microstate is inaccessible when the systems in the ensemble have had ample opportunity to move into it, and yet have not done so. Note that for an equilibrium state, the entropy is just as well-defined as more familiar quantities such as the temperature and the mean pressure.

Consider, again, two systems A and A δ which are in thermal contact but can do no work on one another. Let E and Eδ be the energies of the two systems, and Ω δ (E) and Ω (E) the respective densities of states. Furthermore, let E(0) be the conserved energy of the system as a whole and the Ω(0) corresponding density of states. We have from Eq. that,

$$\Omega(0)\ (E) = \Omega(E)\ \Omega\delta(E\delta),$$

where Eδ = E(0) – E. In other words, the number of states accessible to the whole system is the product of the numbers of states accessible to each subsystem, since every microstate of A can be combined with every microstate of A δ to form a distinct microstate of the whole system that in equilibrium the mean energy of A takes the value Eδ Eδ for which Ω(0)(E) is maximum, and the temperatures of A and Aδ are equal. The distribution of E around the mean value is of order Δ*E= Eδ / f , where f is the number of degrees of freedom. It follows that the total number of accessible microstates is approximately the number of states which lie within Δ*E ofEδ . Thus, the entropy of the whole system is given by,

$$\Omega^*(E) = \frac{\delta^* E}{\delta E}\Omega(E)$$

According to our usual estimate, Ω ~ Ef, the first term on the right-hand side is of order kf whereas the second term is of order k ln (Eδ / f δE). Any reasonable choice for the energy subdivision δE should be greater thanEδ /f,

otherwise there would be less than one microstate per subdivision. It follows that the second term is less than or of order k ln f, which is utterly negligible compared to kf. Thus,

S(0) = k ln W(0) (Eδ) = kln [W(Eδ)W(Eδ ¢)] – k ln W(Eδ) + k ln W¢(Eδ ¢) to an excellent approximation, giving,

$$S(0) = S(E\delta) + S\delta(E\delta).$$

It can be seen that the probability distribution for Ω(0)(E)is so strongly peaked around its maximum value that, for the purpose of calculating the entropy, the total number of states is equal to the maximum number of states [i.e., δ(to0t)]. One consequence of this is that the entropy has the simple additive property Eq. Thus, the total entropy of two thermally interacting systems in equilibrium is the sum of the entropies of each system in isolation.

Uses of Entropy

We have defined a new function called entropy, denoted S, which parameterizes the amount of disorder in a macroscopic system. The entropy of an equilibrium macrostate is related to the number of accessible microstates Ω via,

$$S = k \ln \Omega.$$

On a macroscopic level, the increase in entropy due to a quasi-static change in which an infinitesimal amount of heat dQ is absorbed by the system is given by,

$$dS \;\delta\; dQ\,,$$

where T is the absolute temperature of the system. The second law of thermodynamics states that the entropy of an isolated system can never spontaneously decrease. Let us now briefly examine some consequences of these results.

Consider two bodies, A and Aδ, which are in thermal contact but can do no work on one another. We know what is supposed to happen here. Heat flows from the hotter to the colder of the two bodies until their temperatures are the same. Consider a quasi-static exchange of heat between the two bodies. According to the first law of thermodynamics, if an infinitesimal amount of heat dQ is absorbed by A then infinitesimal heat dQδ= – dQ is absorbed by Aδ. The increase in the entropy of system A is dS = dQ /T and the corresponding increase in the entropy of Aδ is dS = dQ /Tδ. Here, T and Tδ are the temperatures of the two systems, respectively. Note that dQ is assumed to the sufficiently small that the heat transfer does not substantially modify the temperatures of either system. The change in entropy of the whole system is,

$$dS(0)\ dS\ dS\ 1\ 1\ d$$

This change must be positive or zero, according to the second law of thermodynamics, so dS(0) ≥ 0. It follows that dQ is positive (i.e., heat flows from Aδ to A) when Tδ > T, and vice versa. The spontaneous flow of heat only ceases when T = Tδ. Thus, the direction of spontaneous heat flow is a consequence of the second law of thermodynamics.

Note that the spontaneous flow of heat between bodies at different temperatures is always an irreversible process which increases the entropy, or disorder, of the Universe.

Consider, now, the slightly more complicated situation in which the two systems can exchange heat and also do work on one another via a movable partition. Suppose that the total volume is invariant, so that

$$V(0) = V + V\delta \text{ constant,}$$

where V and Vδ are the volumes of A and Aδ, respectively. Consider a quasistatic change in which system A absorbs an infinitesimal amount of heat dQ and its volume simul-taneously increases by an infinitesimal amount dV. The infinitesimal amount of work done by system A is dW = pdV, where p is the mean pressure of A. According to the first law of thermodynamics, dQ δ dE δ dW δ dE δ pdV, where dE is the change in the internal energy of A. Since dS δ dQ/T, the increase in entropy of system A is written,

$$.\ dS\ dE\ pdV$$

Likewise, the increase in entropy of system Aδ is given by, .

dE p dV

According to Eq.

$$\frac{1}{T}=\left(\frac{\partial S}{\partial E}\right)_{v'}$$

$$\frac{\overline{p}}{T}=\left(\frac{\partial S}{\partial E}\right)_{E'}$$

where the subscripts are to remind us what is held constant in the partial derivatives. We can write a similar pair of equations for the system.

The overall system is assumed to be isolated, so conservation of energy gives dE + dEδ= 0. Furthermore, Eq. implies that dV + dVδ= 0. It follows that the total change in entropy is given by,

$$dS^{(0)}=dS+dS'=\left(\frac{1}{T}-\frac{1}{T'}\right)dE+\left(\frac{\overline{p}}{T}-\frac{\overline{p}'}{T'}\right)dV$$

The equilibrium state is the most probable state. According to statistical mechanics, this is equivalent to the state with the largest number of accessible microstates. Finally, Eq. implies that this is the maximum entropy state.

The system can never spontaneously leave a maximum entropy state, since this would imply a spontaneous reduction in entropy, which is forbidden by the second law of thermodynamics.

A maximum or minimum entropy state must satisfy dS(0) = 0 for arbitrary small variations of the energy and external parameters. It follows from Eq. that,

T = T′

$\overline{p}=\overline{p}'$

for such a state. This corresponds to a maximum entropy state (i.e., an equilibrium state) provided,

$$\left(\frac{\partial^2 S}{\partial E^2}\right)_V<0$$

$$\left(\frac{\partial^2 S}{\partial V^2}\right)_E<0$$

with a similar pair of inequalities for system Aδ. The usual estimate Ω δ EfVf, giving S = kf ln E + kf ln V+..., ensures that the above inequalities are satisfied in conventional macroscopic systems.

In the maximum entropy state the systems A and A¢ have equal temperatures (i.e., they are in thermal equilibrium) and equal pressures (i.e., they are in mechanical equilibrium). The second law of thermodynamics implies that the two interacting systems will evolve towards this state, and will then remain in it indefinitely (if left undisturbed).

ENTROPY SYSTEM

Consider an isolated system whose energy is known to lie in a narrow range. Let Ω be the number of accessible microstates. According to the principle of equal a priori probabilities, the system is equally likely to be found in any one of these states when it is in thermal equilibrium. The accessible states are just that set of microstates which are consistent with the macroscopic constraints imposed on the system. These constraints can usually be quantified by specifying the values of some parameters y1..., yn which characterize the macrostate. Note that these parameters are not necessarily external: e.g., we could specify either the volume (an external parameter) or the mean pressure (the mean force conjugate to the volume).

The number of accessible states is clearly a function of the chosen parameters, so we can write Ω δ Ω (y1..., yn)for the number of microstates consistent with a macrostate in which the general parameter yδ lies in the range yδ to yδ + dyδ.

Suppose that we start from a system in thermal equilibrium. According to statistical mechanics, each of the Ωi, say, accessible states are equally likely. Let us now remove, or relax, some of the constraints imposed on the system. Clearly, all of the microstates formally accessible to the system are still accessible, but many additional states will, in general, become accessible. Thus, removing or relaxing constraints can only have the effect of increasing, or possibly leaving unchanged, the number of microstates accessible to the system. If the final number of accessible states is Ωf, then we can write,

$$\Omega f \geq \Omega i.$$

Immediately after the constraints are relaxed, the systems in the ensemble are not in any of the microstates from which they were previously excluded. So the systems only occupy a fraction, of the Ωf states now accessible to them. This is clearly not a equilibrium situation. Indeed, if Ωf >> Ωi then the configuration in which the systems are only distributed over the original Ωi states is an extremely unlikely one. In fact, its probability of occurrence is given by Eq. According to the theorem, H the ensemble will evolve in time until a more probable final state is reached in which the systems are evenly distributed over the Ωf available states.

As a simple example, consider a system consisting of a box divided into two regions of equal volume. Suppose that, initially, one region is filled with gas and the other is empty. The constraint imposed on the system is, thus, that the coordinates of all of the molecules must lie within the filled region.

In other words, the volume accessible to the system V = Vi is, where Vi is half the volume of the box. The constraints imposed on the system can be relaxed by removing the partition and allowing gas to flow into both regions. The volume accessible to the gas is now $V = V_f = 2V_i$. Immediately after the partition is removed, the system is in an extremely improbable state that at constant energy the variation of the number of accessible states of an ideal gas with the volume is,

$$\Omega \ \mu \ VN,$$

where N is the number of particles. Thus, the probability of observing the state immediately after the partition is removed in an ensemble of equilibrium systems with volume $V = V_f$ is,

$$P_i = \frac{\Omega_i}{\Omega_f} = \left(\frac{V_i}{V_f}\right)^N = \left(\frac{1}{2}\right)^N$$

If the box contains of order 1 mole of molecules then N ~ 1024 and this probability is fantastically small:

$$Pi \sim \exp(-10^{24}).$$

Clearly, the system will evolve towards a more probable state.

This can also be phrased in terms of the parameters y1...,yn of the system. Suppose that a constraint is removed. For instance, one of the parameters, y, say, which originally had the value y = yi, is now allowed to vary. According to statistical mechanics, all states accessible to the system are equally likely. So, the probability P(y)of finding the system in equilibrium with the parameter in the range to is just proportional y to y + ∞y the number of microstates in this interval: i.e.,

$$P(y) \infty \ \Omega \ (y).$$

Usually, Ω(y) has a very pronounced maximum at some particular value y∞ .This means that practically all systems in the final equilibrium ensemble have values of y close to y∞ . Thus, if yi ≠ y∞ initially then the parameter y will change until it attains a final value close to y∞ , where Ω is maximum.

If some of the constraints of an isolated system are removed then the parameters of the system tend to readjust themselves in such a way that,

$$\Omega(y1...,yn) \propto \text{maximum}.$$

Suppose that the final equilibrium state has been reached, so that the systems in the ensemble are uniformly distributed over the Ωf accessible final states. If the original constraints are reimposed then the systems in the ensemble still occupy these Ωf states with equal probability. Thus, if $\Omega f > \Omega i$, simply restoring the constraints does not restore the initial situation. Once the systems are randomly distributed over the Ωf states they cannot be expected to spontaneously move out of some of these states and occupy a more restricted class of states merely in response to the reimposition of a constraint. The initial condition can also not be restored by removing further constraints. This could only lead to even more states becoming accessible to the system.

Suppose that some process occurs in which an isolated system goes from some initial configuration to some final configuration. If the final configuration is such that the imposition or removal of constraints cannot by itself restore the initial condition then the process is deemed irreversible. On the other hand, if it is such that the imposition or removal of constraints can restore the initial condition then the process is deemed reversible.

From what we have already said, an irreversible process is clearly one in which the removal of constraints leads to a situation where $\Omega f > \Omega i$. A reversible process corresponds to the special case where the removal of constraints does not change the number of accessible states, so that $\Omega f = \Omega i$. In this situation, the systems remain distributed with equal probability over these states irrespective of whether the constraints are imposed or not.

Our microscopic definition of irreversibility is in accordance with the macroscopic. Recall that on a macroscopic level an irreversible process is one which "looks unphysical" when viewed in reverse. On a microscopic level it is clearly plausible that a system should spontaneously evolve from an improbable to a probable configuration in response to the relaxation of some constraint. However, it is quite clearly implausible that a system should ever spontaneously evolve from a probable to an improbable configuration. Let us consider our example again. If a gas is initially restricted to one half of a box, via a partition, then the flow of gas from one side of the box to the other when the partition is removed is an irreversible process.

This process is irreversible on a microscopic level because the initial configuration cannot be recovered by simply replacing the partition. It is irreversible on a macroscopic level because it is obviously unphysical for the molecules of a gas to spontaneously distribute themselves in such a manner that they only occupy half of the available volume.

It is actually possible to quantify irreversibility. In other words, in addition to stating that a given process is irreversible, we can also give some indication of how irreversible it is. The parameter which measures irreversibility is just the number of accessible states Ω. Thus, if Ω for an isolated system spontaneously increases then the process is irreversible, the degree of irreversibility being proportional to the amount of the increase. If Ω stays the same then the process is reversible. Of course, it is unphysical for Ω to ever spontaneously decrease. In symbols, we can write,

$$\Omega f - \Omega i \propto \Delta\Omega \geq 0,$$

for any physical process operating on an isolated system. In practice, Ω itself is a rather unwieldy parameter with which to measure irreversibility. For instance, in the previous example, where an ideal gas doubles in volume (at constant energy) due to the removal of a partition, the fractional increase in Ω is,

$$\frac{\Omega_f}{\Omega_i} \simeq 10^{2\nu \times 10^{23}}$$

where is the number of moles. This is an extremely large number! It is far more convenient to measure irreversibility in terms of ln Ω. If Eq. is true then it is certainly also true that,

$$In\, \Omega_f - \ln \Omega_i \equiv \Delta \ln \Omega \geq 0$$

for any physical process operating on an isolated system. The increase in when an ideal gas doubles in volume (at constant energy) is,

$$\ln \Omega_f - \ln \Omega_i = \nu N_A \ln 2$$

where $N_A = 6 \times 10^{23}$. This is a far more manageable number! Since we usually deal with particles by the mole in laboratory physics, it makes sense to premultiply our measure of irreversibility by a number of order 1/NA. For historical reasons, the number which is generally used for this purpose is the Boltzmann constant k, which can be written,

$$k = \frac{R}{N_A}\, joules\,/\,kelvin$$

where,

$$R = 8.3143 \text{ joules/kelvin/mole}$$

is the ideal gas constant which appears in the well-known equation of state for an ideal gas, PV = v RT. Thus, the final form for our measure of irreversibility is,

$$S = k \ln \Omega$$

This quantity is termed "entropy", and is measured in joules per degree kelvin. The increase in entropy when an ideal gas doubles in volume (at constant energy) is,

$$Sf - Si = v R \ln 2,$$

which is order unity for laboratory scale systems (i.e., those containing about one mole of particles). The essential irreversibility of macroscopic phenomena can be summed up as follows:

$$Sf - Si = \Delta S \geq 0,$$

for a process acting on an isolated system. Thus, the entropy of an isolated system tends to increase with time and can never decrease. This proposition is known as the second law of thermodynamics.

One way of thinking of the number of accessible states Ω is that it is a measure of the disorder associated with a macrostate. For a system exhibiting a high degree of order we would expect a strong correlation between the motions of the individual particles. For instance, in a fluid there might be a strong tendency for the particles to move in one particular direction, giving rise to an ordered flow of the system in that direction. On the other hand, for a system exhibiting a low degree of order we expect far less correlation between the motions of individual particles.

It follows that, all other things being equal, an ordered system is more constrained than a disordered system, since the former is excluded from microstates in which there is not a strong correlation between individual particle motions, whereas the latter is not. Another way of saying this is that an ordered system has less accessible microstates than a corresponding disordered system. Thus, entropy is effectively a measure of the disorder in a system (the disorder increases with □). With this interpretation, the second law of thermodynamics reduces to the statement that isolated systems tend to become more disordered with time, and can never become more ordered.

Note that the second law of thermodynamics only applies to isolated systems. The entropy of a non-isolated system can decrease. For instance, if a gas expands (at constant energy) to twice its initial volume after the removal of a partition, we can subsequently recompress the gas to its original volume.

The energy of the gas will increase because of the work done on it during compression, but if we absorb some heat from the gas then we can restore it to its initial state. Clearly, in restoring the gas to its original state, we have restored its original entropy.

This appears to violate the second law of thermodynamics because the entropy should have increased in what is obviously an irreversible process (just try to make a gas spontaneously occupy half of its original volume!).

However, if we consider a new system consisting of the gas plus the compression and heat absorption machinery, then it is still true that the entropy of this system (which is assumed to be isolated) must increase in time.

Thus, the entropy of the gas is only kept the same at the expense of increasing the entropy of the rest of the system, and the total entropy is increased. If we consider the system of everything in the Universe, which is certainly an isolated system since there is nothing outside it with which it could interact, then the second law of thermodynamics becomes:

The disorder of the Universe tends to increase with time and can never decrease. An irreversible process is clearly one which increases the disorder of the Universe, whereas a reversible process neither increases nor decreases disorder. This definition is in accordance with our previous definition of an irreversible process as one which "does not look right" when viewed backwards. One easy way of viewing macroscopic events in reverse is to film them, and then play the film backwards through a projector. There is a famous passage in the novel "Slaughterhouse 5," by Kurt Vonnegut, in which the hero, Billy Pilgrim, views a propaganda film of an American World War II bombing raid on a German city in reverse. This is what the film appeared to show:

"American planes, full of holes and wounded men and corpses took off backwards from an airfield in England. Over France, a few German fighter planes flew at them backwards, sucked bullets and shell fragments from some of the planes and crewmen. They did the same for wrecked American bombers on the ground, and those planes flew up backwards and joined the formation.

The formation flew backwards over a German city that was in flames. The bombers opened their bomb bay doors, exerted a miraculous magnetism which shrunk the fires, gathered them into cylindrical steel containers, and lifted the containers into the bellies of the planes. The containers were stored neatly in racks. The Germans had miraculous devices of their own, which were long steel tubes. They used them to suck more fragments from the crewmen and planes. But there were still a few wounded Americans, though, and some of the bombers were in bad repair. Over France, though, German fighters came up again, made everything and everybody as good as new."

Vonnegut's point, I suppose, is that the morality of actions is inverted when you view them in reverse.

What is there about this passage which strikes us as surreal and fantastic? What is there that immediately tells us that the events shown in the film could never happen in reality? It is not so much that the planes appear to fly backwards and the bombs appear to fall upwards. After all, given a little ingenuity and a sufficiently good pilot, it is probably possible to fly a plane backwards.

Likewise, if we were to throw a bomb up in the air with just the right velocity we could, in principle, fix it so that the velocity of the bomb matched that of a passing bomber when their paths intersected. Certainly, if you had never seen a plane before it would not be obvious which way around it was supposed to fly.

However, certain events are depicted in the film, "miraculous" events in Vonnegut's words, which would immediately strike us as the wrong way around even if we had never seen them before. For instance, the film might show thousands of white hot bits of shrapnel approach each other from all directions at great velocity, compressing an explosive gas in the process, which slows them down such that when they meet they fit together exactly to form a metal cylinder enclosing the gases and moving upwards at great velocity.

What strikes us as completely implausible about this event is the spontaneous transition from the disordered motion of the gases and metal fragments to the ordered upward motion of the bomb.

MICROSCOPIC DEFINITION OF ENTROPY (STATISTICAL MECHANICS)

In statistical thermodynamics the entropy is defined as (proportional to) the logarithm of the number of microscopic configurations that result in the observed macroscopic description of the thermodynamic system:

$$S=k_B ln\ \Omega$$

where

kB is Boltzmann's constant 1.38066×10–23 J K–1 and

Ω is the number of microstates corresponding to the observed thermodynamic macrostate.

This definition is considered to be the fundamental definition of entropy. In Boltzmann's 1896 Lectures on Gas Theory, he showed that this expression gives a measure of entropy for systems of atoms and molecules in the gas phase, thus providing a measure for the entropy of classical thermodynamics.

In 1877, Boltzmann visualized a probabilistic way to measure the entropy of an ensemble of ideal gas particles, in which he defined entropy to be proportional to the logarithm of the number of microstates such a gas could occupy. Henceforth, the essential problem in statistical thermodynamics, i.e. according to Erwin Schrödinger, has been to determine the distribution of a given amount of energy E over N identical systems.

Statistical mechanics explains entropy as the amount of uncertainty which remains about a system, after its observable macroscopic properties have been taken into account. For a given set of macroscopic variables, like temperature and volume, the entropy measures the degree to which the probability of the system is spread out over different possible quantum states. The more states available to the system with higher probability, the greater the entropy. More specifically, entropy is a logarithmic measure of the density of states. In essence, the most general interpretation of entropy is as a measure of our uncertainty about a system. The equilibrium state of a system maximizes the entropy because we have lost all information about the initial conditions except for the conserved variables; maximizing the entropy maximizes our ignorance about the details of the system. This uncertainty is not of the everyday subjective kind, but rather the uncertainty inherent to the experimental method and interpretative model.

On the molecular scale, the two definitions match up because adding heat to a system, which increases its classical thermodynamic entropy, also increases the system's thermal fluctuations, so giving an increased lack of information about the exact microscopic state of the system, i.e. an increased statistical mechanical entropy. The interpretative model has a central role in determining entropy. The qualifier "for a given set of macroscopic variables" above has very deep implications: if two observers use different sets of macroscopic variables, then they will observe different entropies. For example, if observer A uses the variables U, V and W, and observer B uses U, V, W, X, then, by changing X, observer B can cause an effect that looks like a violation of the second law of thermodynamics to observer A. In other words: the set of macroscopic variables one chooses must include everything that may change in the experiment, otherwise one might see decreasing entropy.

ENTROPY IN CHEMICAL THERMODYNAMICS

Thermodynamic entropy is central in chemical thermodynamics, enabling changes to be quantified and the outcome of reactions predicted. The second law of thermodynamics states that entropy in the combination of a system and its surroundings increases during all spontaneous chemical and physical processes. Spontaneity in chemistry means "by itself, or without any outside influence", and has nothing to do with speed. The Clausius equation of Θqrev/ T = ΔS introduces the measurement of entropy change, ΔS. Entropy change describes the direction and quantitates the magnitude of simple changes such as heat transfer between systems – always from hotter to cooler spontaneously.

Thus, when a mole of substance at 0 K is warmed by its surroundings to 298 K, the sum of the incremental values of qrev/T constitute each element's or compound's standard molar entropy, a fundamental physical property and an indicator of the amount of energy stored by a substance at 298 K. Entropy change also measures the mixing of substances as a summation of their relative quantities in the final mixture.

Entropy is equally essential in predicting the extent of complex chemical reactions, i.e. whether a process will go as written or proceed in the opposite direction. For such applications, ΔS must be incorporated in an expression that includes both the system and its surroundings, ΔSuniverse =ΔSsurroundings + ΔS system. This expression becomes, via some steps, the Gibbs free energy equation for reactants and products in the system: ΔG [the Gibbs free energy change of the system] = ΔH [the enthalpy change] "T ΔS [the entropy change].

ENTROPY BALANCE EQUATION FOR OPEN SYSTEMS

In chemical engineering, the principles of thermodynamics are commonly applied to "open systems", i.e. those in which heat, work, and mass flow across the system boundary.

In a system in which there are flows of both heat (ΘQ) and work, i.e. Θ WS (shaft work) and P(dV/dt) (pressure-volume work), across the system boundaries, the heat flow, but not the work flow, causes a change in the entropy of the system. This rate of entropy change is Q/T where T is the absolute thermodynamic temperature of the system at the point of the heat flow. If, in addition, there are mass flows across the system boundaries, the total entropy of the system will also change due to this convected flow.

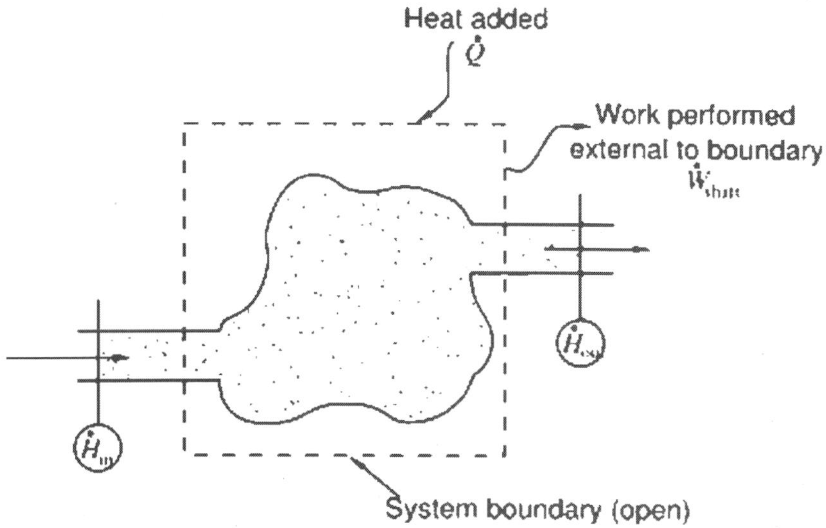

Fig. Entropy Balance Applied to An Open System

To derive a generalized entropy balanced equation, we start with the general balance equation for the change in any extensive quantity Θ in a thermodynamic system, a quantity that may be either conserved, such as energy, or non-conserved, such as entropy. The basic generic balance expression states that dΘ/dt, i.e. the rate of change of Θ in the system, equals the rate at which Θ enters the system at the boundaries, minus the rate at which Θ leaves the system across the system boundaries, plus the rate at which Θ is generated within the system.

Using this generic balance equation, with respect to the rate of change with time of the extensive quantity entropy S, the entropy balance equation for an open thermodynamic system is:

$$\frac{dS}{dt} = \sum_{k=1}^{K} M_k \hat{S}_k + \frac{Q}{T} + S_{gen}$$

M S = the net rate of entropy flow due to the flows of mass into and out of the system (where^S = entropy per unit mass).

Q/T = the rate of entropy flow due to the flow of heat across the system boundary.

ΘSgen = the rate of internal generation of entropy within the system.

Note, also, that if there are multiple heat flows, the term QΘ/T is to be replaced by ΣQ j /Tj, where Qj is the heat flow and Tj is the temperature at the heat flow port into the system.

ENTROPY SYSTEM AND QUANTUM MECHANICS

The entropy of a system is defined in terms of the number Ω of accessible microstates consistent with an overall energy in the range E to E + Evia,

$$S = k \ln \Omega.$$

We have already demonstrated that this definition is utterly insensitive to the resolution δE to which the macroscopic energy is measured. In classical mechanics, if a system possesses f degrees of freedom then phase-space is conventionally subdivided into cells of arbitrarily chosen volume accessible microstates is equivalent to the number of these cells in the volume of phase-space consistent with an overall energy of the system lying in the range E to E +δE. Thus,

$$\Omega = \frac{1}{h_0^{\,f}} \int \ldots \int dq_1 \ldots dq_f dq_1 \ldots dp_f$$

giving,

$$S = k \ln\left(\int \ldots \int dq_1 \ldots dq_f dp_1 \ldots dp_f \right) - kf \ln h_0$$

Thus, in classical mechanics the entropy is undetermined to an arbitrary additive constant which depends on the size of the cells in phase- space. In fact, S increases as the cell size decreases. The second law of thermodynamics is only concerned with changes in entropy, and is, therefore, unaffected by an additive constant. Likewise, macroscopic thermodynamical quantities, such as the temperature and pressure, which can be expressed as partial derivatives of the entropy with respect to various macroscopic parameters are unaffected by such a constant. So, in classical mechanics the entropy is rather like a gravitational potential: it is undetermined to an additive constant, but this does not affect any physical laws. The non-unique value of the entropy comes about because there is no limit to the precision to which the state of a classical system can be specified. In other words, the cell size h0 can be made arbitrarily small, which corresponds to specifying the particle coordinates and momenta to arbitrary accuracy. However, in quantum mechanics the uncertainty principle sets a definite limit to how accurately the particle coordinates and momenta can be specified. In general,

$$\delta q_i \delta p_i \geq h_i$$

where pi is the momentum conjugate to the generalized coordinate qi, and δqi, δpiare the uncertainties in these quantities, respectively. In fact, in quantum mechanics the number of accessible quantum states with the overall energy in the range E to E + δE is completely determined. This implies that, in reality, the entropy of a system has a unique and unambiguous value. Quantum mechanics can often be "mocked up" in classical mechanics by setting the cell size in phase-space equal to Planck's constant, so that h0 = h. This automatically enforces the most restrictive form of the uncertainty principle, δqiδpi = h. In many systems, the substitution h0 δ h in Eq. gives the same, unique value for as that obtained from a full quantum mechanical calculation. Consider a simple quantum mechanical system consisting of N non-interacting spinless particles of mass m confined in a cubic box of dimension L. The energy levels of the th particle are given by,

$$e_i = \frac{\hbar^2 \pi^2}{2mL^2}\left(n_{il}^2 + n_{i2}^2 + n_{i3}^2\right)$$

where ni1, ni2, and ni3 are three (positive) quantum numbers. The overall energy of the system is the sum of the energies of the individual particles, so that for a general state,

$$E_r = \sum_{i=1}^{N} e_i$$

The overall state of the system is completely specified by 3N quantum numbers, so the number of degrees of freedom is f = 3N. The classical limit corresponds to the situation where all of the quantum numbers are much greater than unity.

In this limit, the number of accessible states varies with energy according to our usual estimate Ω δ Ef.

The lowest possible energy state of the system, the so-called ground-state, corresponds to the situation where all quantum numbers take their lowest possible value, unity. Thus, the ground-state energy E0 is given by,

$$E_0 = \frac{f\hbar^2 \pi^2}{2mL^2}$$

There is only one accessible microstate at the ground-state energy (i.e., that where all quantum numbers are unity), so by our usual definition of entropy,

$$S(E_0) = kln1 = 0$$

In other words, there is no disorder in the system when all the particles are in their ground-states.

Clearly, as the energy approaches the ground-state energy, the number of accessible states becomes far less than the usual classical estimate Ef. This is true for all quantum mechanical systems. In general, the number of microstates varies roughly like,

$$\Omega(E) \sim 1 + C(E - E_0)^f$$

where C is a positive constant. According to Eq. the temperature varies approximately like,

$$T \sim \frac{E - E_0}{kf}$$

provided Ω >> 1. Thus, as the absolute temperature of a system approaches zero, the internal energy approaches a limiting value E0(the quantum mechanical ground-state energy), and the entropy approaches the limiting value zero. This proposition is known as the third law of thermodynamics.

At low temperatures, great care must be taken to ensure that equilibrium thermodynamical arguments are applicable, since the rate of attaining equilibrium may be very slow. Another difficulty arises when dealing with a system in which the atoms possess nuclear spins.

Typically, when such a system is brought to a very low temperature the entropy associated with the degrees of freedom not involving nuclear spins becomes negligible.

Nevertheless, the number of microstates Ωs corresponding to the possible nuclear spin orientations may be very large. Indeed, it may be just as large as the number of states at room temperature.

The reason for this is that nuclear magnetic moments are extremely small, and, therefore, have extremely weak mutual interactions. Thus, it only takes a tiny amount of heat energy in the system to completely randomize the spin orientations. Typically, a temperature as small as 10–3 degrees kelvin above absolute zero is sufficient to randomize the spins.

Suppose that the system consists of N atoms of spin 1/2. Each spin can have two possible orientations. If there is enough residual heat energy in the system to randomize the spins then each orientation is equally likely. If follows that there are Ωs= 2N accessible spin states. The entropy associated with these states is S0 = k ln Ωs = v R ln 2. Below some critical temperature, T0, the interaction between the nuclear spins becomes significant, and the system settles down in some unique quantum mechanical ground-state (e.g., with all spins aligned).

In this situation, S→0, in accordance with the third law of thermodynamics. However, for temperatures which are small, but not small enough to "freeze out" the nuclear spin degrees of freedom, the entropy approaches a limiting value S0 which depends only on the kinds of atomic nuclei in the system. This limiting value is independent of the spatial arrangement of the atoms, or the interactions between them. Thus, for most practical purposes the third law of thermodynamics can be written,

as $T \to 0_+, S \to S_0$

where 0+denotes a temperature which is very close to absolute zero, but still much larger than. This modification of the third law is useful because it can be applied at temperatures which are not prohibitively low.

ENTROPY AND INFORMATION THEORY

In information theory, entropy is the measure of the amount of information that is missing before reception and is sometimes referred to as Shannon entropy. Shannon entropy is a broad and general concept which finds applications in information theory as well as thermodynamics. It was originally devised by Claude Shannon in 1948 to study the amount of information in a transmitted message. The definition of the information entropy is, however, quite general, and is expressed in terms of a discrete set of probabilities pi. In the case of transmitted messages, these probabilities were the probabilities that a particular message was actually transmitted, and the entropy of the message system was a measure of how much information was in the message.

For the case of equal probabilities (i.e. each message is equally probable), the Shannon entropy (in bits) is just the number of yes/no questions needed to determine the content of the message.

The question of the link between information entropy and thermodynamic entropy is a hotly debated topic. Some authors argue that there is a link between the two, while others will argue that they have absolutely nothing to do with each other.

The expressions for the two entropies are very similar. The information entropy H for equal probabilities pi = p is:

$$H = K \ln (1/p)$$

where K is a constant which determines the units of entropy. For example, if the units are bits, then K=1/ln. The thermodynamic entropy S , from a statistical mechanical point of view was first expressed by Boltzmann:

$$S = 1 \ln (1/p)$$

where p is the probability of a system being in a particular microstate, given that it is in a particular macrostate, and k is Boltzmann's constant. It can be seen that one may think of the thermodynamic entropy as Boltzmann's constant, divided by ln, times the number of yes/no questions that must be asked in order to determine the microstate of the system, given that we know the macrostate.

The link between thermodynamic and information entropy was developed in a series of papers by Edwin Jaynes beginning in 1957.

The problem with linking thermodynamic entropy to information entropy is that in information entropy the entire body of thermodynamics which deals with the physical nature of entropy is missing. The second law of thermodynamics which governs the behaviour of thermodynamic systems in equilibrium, and the first law which expresses heat energy as the product of temperature and entropy are physical concepts rather than informational concepts.

If thermodynamic entropy is seen as including all of the physical dynamics of entropy as well as the equilibrium statistical aspects, then information entropy gives only part of the description of thermodynamic entropy. Some authors, like Tom Schneider, argue for dropping the word entropy for the H function of information theory and using Shannon's other term "uncertainty" instead.

Ice Melting Example

The illustration for this article is a classic example in which entropy increases in a small 'universe', a thermodynamic system consisting of the 'surroundings' (the warm room) and 'system' (glass, ice, cold water). In this universe, some heat energy δQ from the warmer room surroundings (at 298 K or 25 °C) will spread out to the cooler system of ice and water at its constant temperature T of 273 K (0°C), the melting temperature of ice.

The entropy of the system will change by the amount $dS = \delta Q/T$, in this example $\delta Q/273$ K. (The heat δQ for this process is the energy required to change water from the solid state to the liquid state, and is called the enthalpy of fusion, i.e. the ΔH for ice fusion.)

The entropy of the surroundings will change by an amount dS = "$\delta Q/298$ K. So in this example, the entropy of the system increases, whereas the entropy of the surroundings decreases.

It is important to realize that the decrease in the entropy of the surrounding room is less than the increase in the entropy of the ice and water: the room temperature of 298 K is larger than 273 K and therefore the ratio, (entropy change), of $\delta Q/298$ K for the surroundings is smaller than the ratio (entropy change), of $\delta Q/273$ K for the ice+water system.

To find the entropy change of our "universe", we add up the entropy changes for its constituents: the surrounding room and the ice+water. The total entropy change is positive; this is always true in spontaneous events in a thermodynamic system and it shows the predictive importance of entropy: the final net entropy after such an event is always greater than was the initial entropy.

As the temperature of the cool water rises to that of the room and the room further cools imperceptibly, the sum of the $\delta Q/T$ over the continuous range, at many increments, in the initially cool to finally warm water can be found by calculus. The entire miniature "universe", i.e. this thermodynamic system, has increased in entropy. Energy has spontaneously become more dispersed and spread out in that "universe" than when the glass of ice water was introduced and became a "system" within it.

Entropy and Cosmology

As a finite universe may be considered an isolated system, it may be subject to the Second Law of Thermodynamics, so that its total entropy is constantly increasing. It has been speculated that the universe is fated to a heat death in which all the energy ends up as a homogeneous distribution of thermal energy, so that no more work can be extracted from any source.

If the universe can be considered to have generally increasing entropy, then – as Roger Penrose has pointed out – gravity plays an important role in the increase because gravity causes dispersed matter to accumulate into stars, which collapse eventually into black holes.

Jacob Bekenstein and Stephen Hawking have shown that black holes have the maximum possible entropy of any object of equal size. This makes them likely end points of all entropy-increasing processes, if they are totally effective matter and energy traps. Hawking has, however, recently changed his stance on this aspect.

The role of entropy in cosmology remains a controversial subject. Recent work has cast extensive doubt on the heat death hypothesis and the applicability of any simple thermodynamic model to the universe in general.

Although entropy does increase in the model of an expanding universe, the maximum possible entropy rises much more rapidly – thus entropy density is decreasing with time. This results in an "entropy gap" pushing the system further away from equilibrium. Other complicating factors, such as the energy density of the vacuum and macroscopic quantum effects, are difficult to reconcile with thermo-dynamical models, making any predictions of largescale thermodynamics extremely difficult.

IRREVERSIBLE PROCESSES

In order that a process may be strictly reversible, it is necessary that the state of the working substance should be one of equilibrium at uniform pressure and temperature throughout.

If heat passes " of itself " from a higher to a lower temperature by conduction, convection or radiation, the transfer cannot be reversed without an expenditure of work. If mechanical work or kinetic energy is directly converted into heat by friction, reversal of the motion does not restore the energy so converted. In all such cases there is necessarily, by Carnot's principle, a loss of efficiency or available energy, accompanied by an increase of entropy, which serves as a convenient measure or criterion of the loss.

A common illustration of an irreversible process is the expansion of a gas into a vacuum or against a pressure less than its own. In this case the work of expansion, pdv, is expended in the first instance in producing kinetic energy of motion of parts of the gas. If this could be co-ordinated and utilized without dissipation, the gas might conceivably be restored to its initial state; but in practice violent local differences of pressure and temperature are produced, the kinetic energy is rapidly converted into heat by viscous eddy friction, and residual differences of temperature are equalized by diffusion throughout the mass.

Even if the expansion is adiabatic, in the sense that it takes place inside a non-conducting enclosure and no heat is supplied from external sources, it will not be isentropic, since the heat supplied by internal friction must be included in reckoning the change of entropy. Assuming that no heat is supplied from external sources and no external work is done, the intrinsic energy remains constant by the first law. The final state of the substance, when equilibrium has been restored, may be deduced from this condition, if the energy can be expressed in terms of the co-ordinates.

But the line of constant energy on – the diagram does not represent the path of the transformation, unless it be supposed to be effected in a series of infinitesimal steps between each of which the substance is restored to an equilibrium state. An irreversible process which permits a more complete experimental investigation is the steady flow of a fluid in a tube.

If the tube is a perfect non-conductor, and if there are no eddies or frictional dissipation, the state of the substance at any point of the tube as to E, p, and v, is represented by the adiabatic or isentropic path, dE= -pdv. The tube varies, the change of kinetic energy of flow, dU, is represented by The flow in this case is reversible, and the state of the fluid is the same at points where the section of the tube is the same.

In practice, however, there is always some frictional dissipation, accompanied by an increase of entropy and by a fall of pressure. In the limiting case of a long fine tube, the bore of which varies in such a manner that U is constant, the state of the substance along a line of flow may be represented by the line of constant total heat, d(E+pv) = o; but in the case of a porous plug or small throttling aperture, the steps of the process cannot be followed, though the final state is the same.

DEVELOPMENT OF QUANTUM MECHANICS

Quantum mechanics is a general theory. It is presumed to apply to everything, from subatomic particles to galaxies. But interest is naturally focused on those phenomena that are most distinctive of quantum mechanics, some of which led to its discovery. Rather than retelling the historical development of quantum theory. The beautiful mathematical development of Newton's mechanics, coupled with increasingly sophisticated technology, predicted the movements of the solar system to incredible accuracy, apart from a tiny discrepancy in the orbit of Mercury. It had been less than a hundred years since it was realised that an electric current could exert a force on a magnet, but that discovery had led to power stations, electric trains and a network of telegraph wires across land and under the oceans.

It had also been only a hundred years since it had been established that light was a wave, and only forty years since Maxwell's realization that the waves in a light signal were electric and magnetic fields, satisfying a wave equation he was able to derive purely by considering electric and magnetic field phenomena. In particular, he was able to predict the speed of light by measuring the electrostatic attractive forces between charges and the magnetic forces between currents.

At about the same time, in the 1860's, Maxwell and Boltzmann gave a brilliant account of the properties of gases by assuming that they were made up of weakly interacting molecules flying about in a container, bouncing off the sides, with a statistical distribution of energies so that the probability of a molecule having energy E was proportional to e-E/kT, k being a universal constant known as Boltzmann's constant.

Boltzmann generalized this result from a box of gas to any system. For example, a solid can be envisioned classically as a lattice of balls (the atoms) connected by springs, which can sustain oscillations in many different ways, each such mode can be thought of as a simple harmonic oscillator, with reasonable approximations concerning the properties of the springs, etc. Boltzmann's work leads to the conclusion that each such mode of oscillation, or degree of freedom, would at temperature T have average energy kT, made up of ½kT potential energy, ½ kT kinetic energy. Notice that this average energy is independent of the strength of the springs, or the masses! All modes of vibration, which will vibrate at very different rates, contain the same energy at the same temperature. This equal sharing is called the Equipartition of Energy. It is not difficult to check this for a one-dimensional classical harmonic oscillator, averaging the energy by integrating over all displacements and momenta (independently) with the weighting factor e-E/kT (which of course needs to be normalized). The result doesn't depend on the spring constant or the mass. Boltzmann's result gave an excellent account of the specific heats of a wide range of materials over a wide temperature range, but there were some exceptions, for example hydrogen gas at low temperatures, and even solids at low enough temperatures. Still, it was generally felt these problems could be handled within the existing framework, just as the slightly odd behaviour of Mercury was likely caused by a small planet, named Vulcan, closer to the sun, and so very hard to observe.

BLACK BODY RADIATION

Black-body radiation is the type of electromagnetic radiation within or surrounding a body in thermodynamic equilibrium with its environment, or emitted by a black body (an opaque and non-reflective body) held at constant, uniform temperature. The radiation has a specific spectrum and intensity that depends only on the temperature of the body. But there was one problem that was hard to get a grip on, an apparently blatant violation of the equipartition of energy. Consider an oven with a small hole in the door, through which the radiation inside is observed. This oven can be heated until it's white hot.

The radiation inside is infrared at low temperatures, becoming visible light as the temperature increases. So, the oven's full of electromagnetic waves, satisfying Maxwell's wave equation, with boundary conditions at the walls of the oven, the electric field has to be essentially zero there, because the walls conduct currents. Of course, the radiation originates in oscillating charges in the walls, using the same analysis of Maxwell's equations that gives the radiation form an antenna. Anyway, there is a set of standing wave modes of electromagnetic vibrations inside the oven, just a three-dimensional version of the series of allowed standing wave modes of vibration of a string fixed at both ends.

So, we should be able to find the energy density of these waves using the same ideas that worked pretty well for the specific heats of solids and gases, that is to say, assume there's kT of energy in each mode of vibration. (This is ½kT of kinetic energy, ½kT of potential energy for each independent direction of vibration.) But this leads to disaster. The problem is that there are infinitely many modes of vibration of the electromagnetic field in an oven.

There is no upper limit to the number of wiggles the wave can have between the walls. So, if we take kT in each mode, we deduce that the oven contains an infinite amount of energy, and radiates an infinite amount through our small hole. This analysis gives no clue as to why the colour we see changes with temperature. Evidently, equipartition of energy isn't working in this case. There's only a finite amount of energy in the oven and at low temperatures there's no energy at all in the modes corresponding to visible light, although that changes as things get hotter.

In the 1890's, German experimentalists measured the energy density as a function of wavelength to great precision, it's called the black body radiation spectrum. A theorist, Planck, found a mathematical formula that fitted this curve exactly,

$$R_T(f)\,df = \frac{8\pi Vhf^3 df / c^3}{e^{hf/kT} - 1}$$

He did not at first have any theoretical justification for this formula, but it was a very accurate fit to some very precise experiments for a suitable value of the constant h, which we discuss in a moment. Factoring out the number of modes of oscillation in the frequency range df, Planck's formula gives the average energy per mode to be,

$$\frac{hf}{e^{hf/kT} - 1}$$

For low frequencies, hf << kT, this correctly gives kT per mode. But, for higher frequencies it's clear that the oscillators are not getting their "fair share" kT of energy. Somehow, the oscillating charges in the walls are not radiating so much energy at the high frequencies. The only way Planck could derive the formula theoretically was by making a weird assumption: he assumed that the oscillating charges in the walls could not just radiate energy continuously, as Maxwell's equations would predict (and as was known to be true for ordinary antennas) but were only allowed to radiate energy in chunks he called quanta.

Furthermore, the amount of energy in one quantum depended on the frequency of the oscillation, in fact linearly: for frequency f, the quantum has energy hf, where h is the constant introduced into the formula above, now known as Planck's constant. It follows that the oscillators themselves could only be oscillating with energies that form a ladder with steps hf apart, above some lowest energy which would be their energy at absolute zero temperature.

The formula follows if we assume the oscillating field component in the oven having frequency f can only have a whole number of quanta of energy, that is to say, its energy must be one of: 0, hf, 2hf, 3hf. The relative probability of it having energy E is e–E/kT, then its relative probabilities of having energy 0, hf, 2hf, are in the ratio 1: e–hf/kT: e–2hf/kT, etc.

The actual probabilities are given by dividing these relative probabilities by the sum of all of them. They clearly are the terms of a geometric series, so their sum is just 1/(1 - e-hf/kT). So, to find the average energy in the oscillator, we take the possible energies 0, hf, 2hf, 3hf, and weight each of them with their probability of occurring, that is, we must find,

$$0.1 + hf.e^{-hf/kT} + 2hf.e^{-2hf/kT} + \ldots$$

and divide the sum by 1/(1 - e-hf/kT).

So, Planck's quantum assumption explains the observed black body radiation curve. It also gives a qualitative explanation of the change in colour of the radiated light as the temperature is increased. The oscillators in the walls derive their energy from the heat vibrations of neighbouring molecules: typically, such a vibration has energy of order kT, with probabilities of more energy going down as e-E/kT.

This means that if the potentially radiating oscillator can only absorb energies in quanta hf, if kT << hf, it will be very unlikely to absorb any energy, and therefore very unlikely to radiate. In the three-dimensional oven, the number of standing wave oscillations in a small frequency range Δf increases with f as f2, so we find that the maximum radiation intensity occurs at a frequency f such that hf is of order kT.

Therefore, as the temperature increases, the frequency at which the most intense radiation occurs increases, and hence the colour moves from red to blue.

The Photoelectric Result

If light shines on certain metals, electrons are emitted. This is the photoelectric effect. If the metal is in air, the electrons bounce off air molecules and are almost certainly rapidly reabsorbed, but if the metal surface is in a vacuum, the electrons can fly away, and in a vacuum tube they can be collected by another piece of metal, and light can cause a current to flow, the origin of the photoelectric cell. In 1902, Lenard studied how the energy of the emitted photoelectrons varied with the intensity of the light. He used a carbon arc light, and could increase the intensity a thousand-fold.

The ejected electrons hit another metal plate, the collector, which was connected to the cathode by a wire with a sensitive ammeter, to measure the current produced by the illumination. To measure the energy of the ejected electrons, Lenard charged the collector plate negatively, to repel the electrons coming towards it.

Thus, only electrons ejected with enough kinetic energy to get up this potential hill would contribute to the current. Lenard discovered that there was a well-defined minimum voltage that stopped any electrons getting through, we'll call it V_{stop}.

To his surprise, he found that Vstop did not depend at all on the intensity of the light! Doubling the light intensity doubled the number of electrons emitted, but did not affect the energies of the emitted electrons. He also discovered, by using light of different colours, that the maximum electron energy did increase as the frequency of the incident light increased.

EINSTEIN SUGGESTS AND CLARIFICATION

In 1905 Einstein gave a very simple interpretation of Lenard's results. He just assumed that the incoming radiation should be thought of as quanta of frequency hf, with f the frequency. In photoemission, one such quantum is absorbed by one electron. If the electron is some distance into the material of the cathode, some energy will be lost as it moves towards the surface. There will always be some electrostatic cost as the electron leaves the surface, this is usually called the work function, W.

The most energetic electrons emitted will be those very close to the surface, and they will leave the cathode with kinetic energy,

$$E = hf - W.$$

On cranking up the negative voltage on the collector plate until the current just stops, that is, to Vstop, the highest kinetic energy electrons must have had energy eVstop on leaving the cathode.

Thus,

$$eV_{stop} = hf - W$$

Thus Einstein's theory makes a very definite quantitative prediction: if the frequency of the incident light is varied, and Vstop plotted as a function of frequency, the slope of the line should be h/e. It is also clear that there is a minimum light frequency for a given metal, that for which the quantum of energy is equal to the work function. Light below that frequency, no matter how bright, will not cause photoemission.

MILLIKAN'S ATTEMPTS TO DISPROVE EINSTEIN'S THEORY

Albert Einstein's suspicions were well justified, for he had spent a lifetime trying (and failing) to create a unified field theory of matter that explained both Quantum Theory/ Light and Relativity/ Gravity. If we accept Einstein's theory, then, this is a completely different way to measure Planck's constant.

The American experimental physicist Robert Millikan, who did not accept Einstein's theory, which he saw as an attack on the wave theory of light, worked for ten years, until 1916, on the photoelectric effect, to disprove Einstein's theory.

He even devised techniques for scraping clean the metal surfaces inside the vacuum tube. For all his efforts he found disappointing results: Einstein's theory, measuring Planck's constant to within 0.5per cent by this method.

One consolation was that he did get a Nobel prize for this series of experiments.

The point to be emphasized is that the same value for Planck's constant,

$$6.6 \times 10^{-34}$$

Joule.sec, emerges from two completely different experiments: the measurement of black body radiation, and measuring energies of emitted electrons in the photoelectric effect.

This is clearly a general property of electromagnetic radiation, and is confirmed by many later experiments, for example Compton scattering, in which light scatters off electrons. By measuring the energy change and momentum change of the electron, it is found that a single quantum of light was scattered.

Hole Argument and Entwurf Theory

While developing general relativity, Einstein became confused about the gauge invariance in the theory. He formulated an argument that led him to conclude that a general relativistic field theory is impossible. He gave up looking for fully generally covariant tensor equations, and searched for equations that would be invariant under general linear transformations only. In June, 1913 the Entwurf ("draft") theory was the result of these investigations. As its name suggests, it was a sketch of a theory, with the equations of motion supplemented by additional gauge fixing conditions. Simultaneously less elegant and more difficult than general relativity, after more than two years of intensive work Einstein abandoned the theory in November, 1915 after realizing that the hole argument was mistaken.

❐

Index

❑❑❑